TOOL MAINTENANCE AND REPAIR GUIDE

By Brent Thompson

Chapters

Drills
Saws
Sanders

3. **Tool Repair**

In General Repair
Compressors Electric
  Pressure Switch
  Motor
  Discharge Manifold
Compressors Gasoline Engine
  Pilot Valve
  Engine
Generators
Nail, Brad, and Staple Guns
Drills
  AC
  DC
Saws
Sanders
Rivet gun (air/hydraulic)
Impact (air and elect)

4. **Tools In General**
    Tips on buying tools

5. **Safety**

Introduction

I have been repairing tools for about ten years. I have worked in a tool repair shop for five years. I love the challenge and working with my hands. I have always repaired my own tools.

I have seen a lot of sites on the Internet where people are asking questions about almost any and every tool. I decided I wanted to put what I have learned into words for those people to be able to work on their tools and save money. I am just a regular guy, no engineer, and will explain these repairs in a way I hope you will be able to follow.

When I am working on tools for someone else, I am trying to accomplish a number of things. I want to treat people the way I want to be treated. When people take the time to bring me a tool I want to not only fix the problem, but check it out completely for other possible problems.

There are two groups of people who want their tools repaired. The first is the person using his tools to make a living. The second is the homeowner using his tools for a hobby or home improvement projects.

I would like to help the guy that uses his tools to make a living have less down time, save money, and make his tools last longer. I want to keep him working. I want to help the homeowner save money, repair his own tools, and maintain his tools.

I do not know how to repair every tool. I will address repairing tools I do know how to repair and the equipment used in that repair. There are so many different tools and variations that it will be hard to cover all tools specifically. The tips will cover what to do in most cases and will cover the basics for most tools. For repairs, you will have to use what information I give and apply what is applicable to your tool.

All tool repair starts with a schematic of that tool. Unless you have repaired this tool before you need the schematic. The schematic shows how the tool comes apart, what parts are in it and part numbers for ordering.

The pictures I have used in this book are of my own tools or I own what is in the picture. I hope they are good enough to help you understand the function of that tool.

Chapter 1

Equipment Used to Repair Tools

Volt Ohm Meter (VOM)

The volt ohm meter will be used to diagnose a number of electrical problems. It will be used to show volts ac, volts dc, amps ac, amps dc, and ohms or resistance (continuity). Ohms is resistance in the flow of electricity and is also known as checking the continuity. Continuity means that you have a closed circuit or a continuous loop with no breaks in the wire or component. On some VOM's there is a diode and capacitor check. You can check a diode and capacitor with the ohm setting of the vom. This is to see if there is a break in the path the electrical current is following.

I will explain the symbols on the meter selector dial. The V with the wavy line over it is Volts AC. The V with a solid line and dotted line over it is Volts DC. The A with the wavy line over it is Amps AC. The A with the straight and dotted line above it is Amps DC. The horse shoe looking symbol is the Ohm resistance or continuity. The arrowhead looking symbol is the diode check selection. The symbol above it is the capacitor check setting. The black lead will always go in the com (common) socket. The red lead has two sockets you can use. The socket that will be used mostly is the socket with the V, horseshoe, and miliamps. This is for voltage both AC and DC, ohms or continuity, and miliamps. The socket above it is 10A. This is to test amps less than 10. The only use for this is very, very low current DC. The little m is miliamp. The crazy looking u is microfares. This is a reading used with capacitors.

If you have a VOM, get it out and take a look at the selector settings. If you want to check AC electrical current you have to know what current you are looking for. You have to select the highest current number on the scale so your test current will be under that. If you are checking for 120 volts AC set the selector to 200 volts AC. If you don't have the higher number selected it may blow a fuse in the VOM or burn it up. This will be true for DC as well. AC is very dangerous, so be very careful. This current is dangerous and can grab you and won't let go. Once you make your check be sure to unplug the tool so you won't get shocked when working on it.

The DC volt and amp readings are not as dangerous as the AC volts. You should still dial in that higher number on the VOM selector though.

The two probes or leads are black and red. The black probe is plugged into the com (common) socket. There are two sockets for the red probe. The one used the most will be ACV and Ohms. The other is for DC volts and amps.

If you want to check an 18v battery to see if it is charged, make sure the red lead is plugged in to the DCV socket. Put your selector switch on a number higher than 18v. Touch the black lead to the negative post and the red to the positive post. On a digital meter you will see what the voltage is on the readout, but on an analog meter you will have to check the needle on the proper scale to get your reading. I like the digital meter better. Some digital meters have automatic settings. When you set the dial to volts you can adjust the decimal, but it will automatically read up to the max voltage.

To check for a broken wire in your cord, you will select the ohm setting and have the red lead in the ohm socket. This setting will be used more than the others in diagnosing electrical problems.

Some VOM's have a diode check selector setting. This will show if the diode is good or broken. It will also show what the current flow is. A diode allows current to move in one direction. It is still basically a wire. We are only concerned about the diode being good or bad in these repairs. The diode setting is telling you the amount of current flow through the diode. The Ohm reading tells you if the wire is broken or not. There are times when you may need to know how much current is flowing through the diode, but not with these repairs.

Some VOM's have a capacitor check selector setting. A capacitor holds a charge, so don't touch the posts. This charge will shock you badly. Take a screw driver, holding the handle, touch the two leads with the blade to discharge the capacitor. This tool will show if the capacitor is good and what the charge is. Here we are concerned about it being good or bad. You can also check a capacitor with the ohm setting. I should mention that there is a capacitor testing tool that may be more accurate because it charges the capacitor and discharges it. This tool will make a certain sound that tells you if the capacitor is good or bad.

Bearing Puller or Splitter

This tool will be used to remove bearings from a shaft. They come in different sizes. If you have tools with small bearings to medium size you can purchase one size. They run from $20 each to hundreds of dollars for a set. The puller is in two halves with a hole in the center when closed for the tool shaft to move through. There are two bolts that can be adjusted on the sides to move the wedges in under the bearing. Some times this alone will dislodge the bearing. If not, you will have to use an arbor press to push the shaft through the puller.

Don't use a hammer because it will flare the top of the shaft. Always push the shaft out by using the arbor press and if necessary use a bolt or socket to aid in pushing. This protects the bearing. If you have to use a hammer, put a piece of wood between the hammer and the shaft.

Wheel Puller

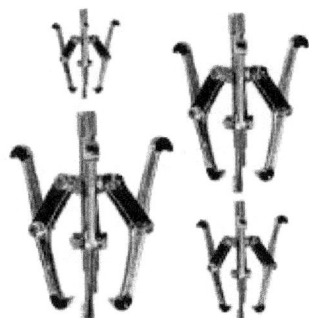

This tool may not be used much, but is handy when needed. It has three fingers with a long bolt going through the middle. If you have a pulley or gear on a shaft, this tool could be used to remove it. These also come in different sizes. The fingers will lock behind the pulley and the end of the bolt will meet the end of the shaft. When you screw the bolt down it pulls the pulley off. When using these tools always make sure you protect the item if you are going to reuse it. You also need to protect it when installing a new one. Look around and use a large washer, tool socket, short piece of pipe or whatever you can find that will protect the item.

Arbor press

This press uses gears (#5) and leverage (#4) to press a bearing on to a shaft. It can also be used to press the shaft through the bearing (#3) and bearing puller. You will need to use something like a large washer, a tool socket, or something similar to protect the bearing. You want to protect the bearing putting it on the shaft so it will still function. Taking off a bad bearing that is damaged isn't a big deal. It is a big deal if you need to reuse it. #2 is used to hold the shaft while you press. You can also use the solid part for pressing bearings on. I modified mine so I can take this part out and get a longer shaft under the press.

The above picture shows how you block up the armature in order to push the shaft through the bearing using the bearing puller and arbor press.

In the above picture you can see how I use the socket and large washer to protect the bearing when pushing it on.

The above picture shows how the bearing puller looks under the bearing and above the fan. The trick is getting the puller under the arbor press and not damage the fan when you push the shaft through.

Allen Wrenches and Star Bits

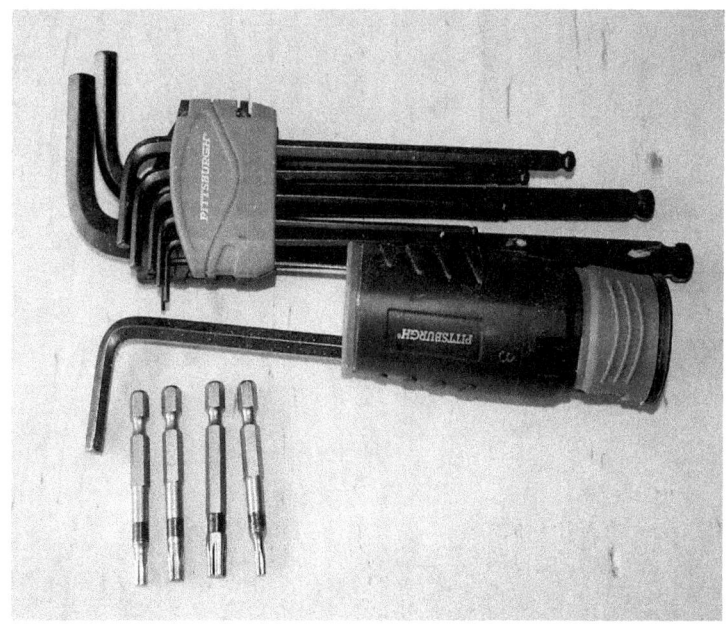

Allen wrenches and star bits will be used a lot on nail guns, drills, and a lot of other tools. Some allen wrenches are made to be used with sockets for those very tight cap screws. The other allen wrenches are hand held, some short, and some long. You can see these in the picture have a ball on the end to allow some movement of the shaft and still be able to turn the screw. There is also a handle provided to help with tight cap screws.

The star bits and philips bits are used with most plastic bodied tools. The best ones to use are the longer ones and hardened. If they are not hardened they will strip out if the screw is too tight.

You can see the star bits are for a quick connect chuck. I use these and other bits in my impact to remove the screw quicker.

Impact Gun

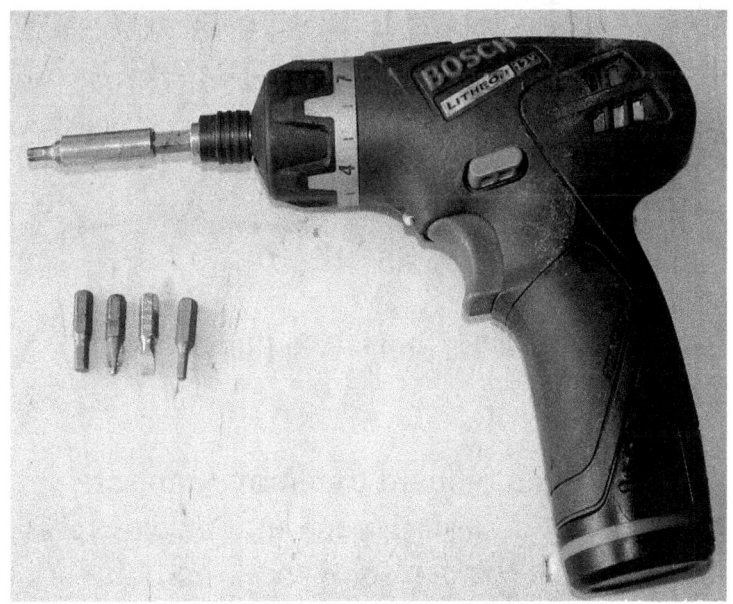

Chisels and Easy Outs

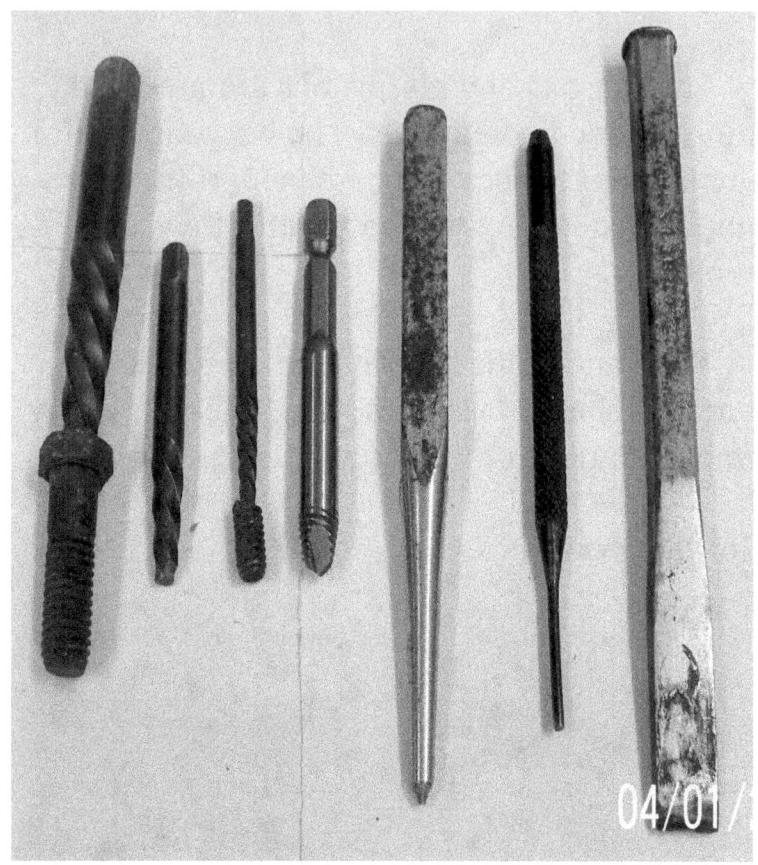

From left four easy outs, two punches, and chisel

I use a chisel on hard to remove cap screws that the socket has stripped out. Place the chisel on the side of the cap screw head and hit with a hammer. After doing this several times it should loosen the screw. Sometimes you can hammer a star bit into the top of the cap screw and get it to turn.

Easy outs can also be used on cap screws or broken bolts. For other screws and broken bolts you may need to drill a hole to use the easy out. I try to have square easy outs so that if I have a left handed thread screw to get out I can. With the square easy out it will turn either direction.

Schematics

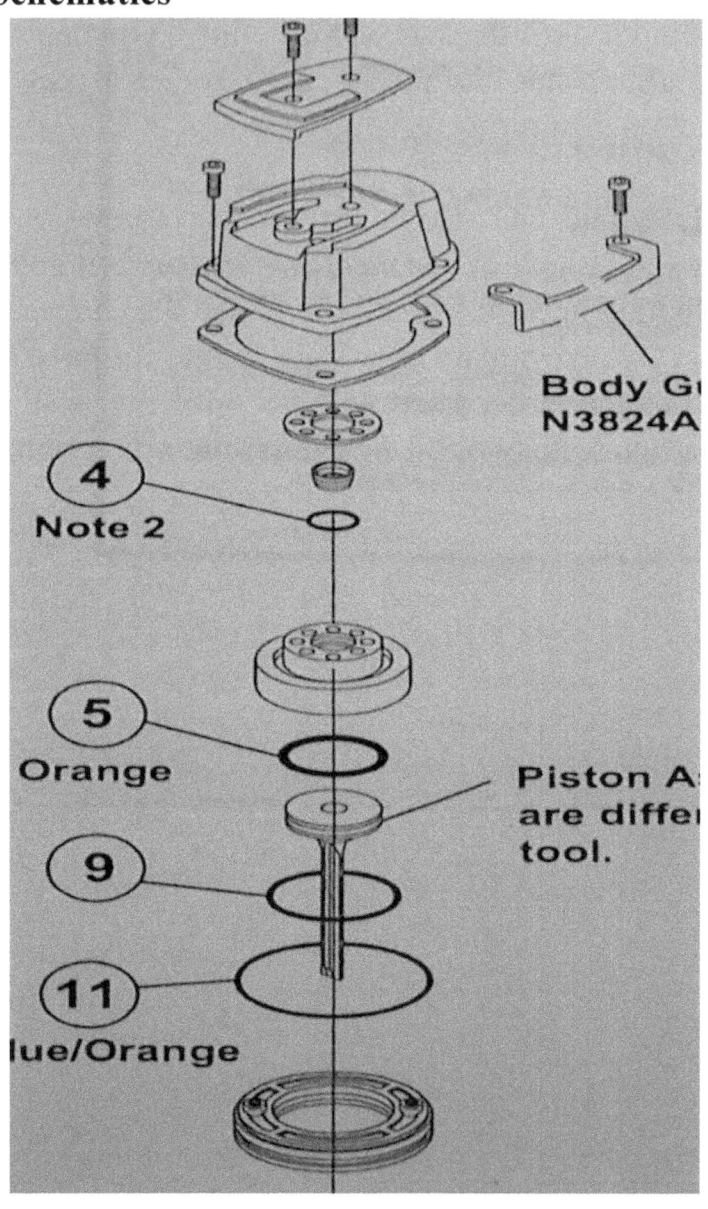

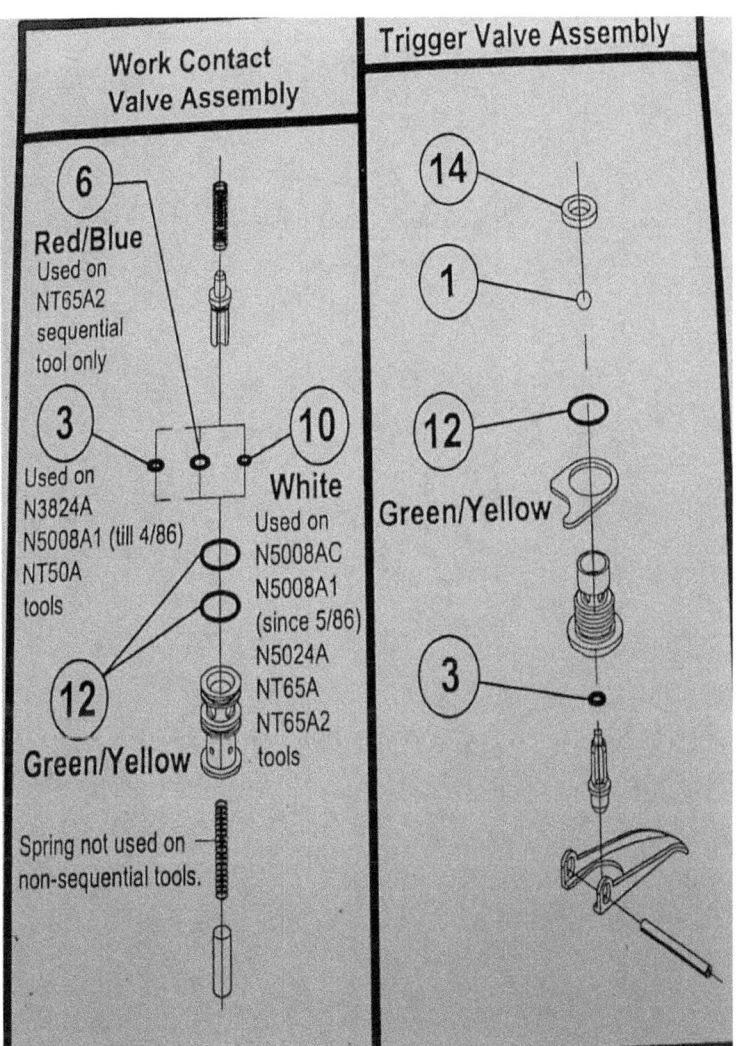

Work Contact Valve Assembly

6 Red/Blue
Used on NT65A2 sequential tool only

3 Used on N3824A N5008A1 (till 4/86) NT50A tools

10 White
Used on N5008AC N5008A1 (since 5/86) N5024A NT65A NT65A2 tools

12 Green/Yellow

Spring not used on non-sequential tools.

Trigger Valve Assembly

14

1

12 Green/Yellow

3

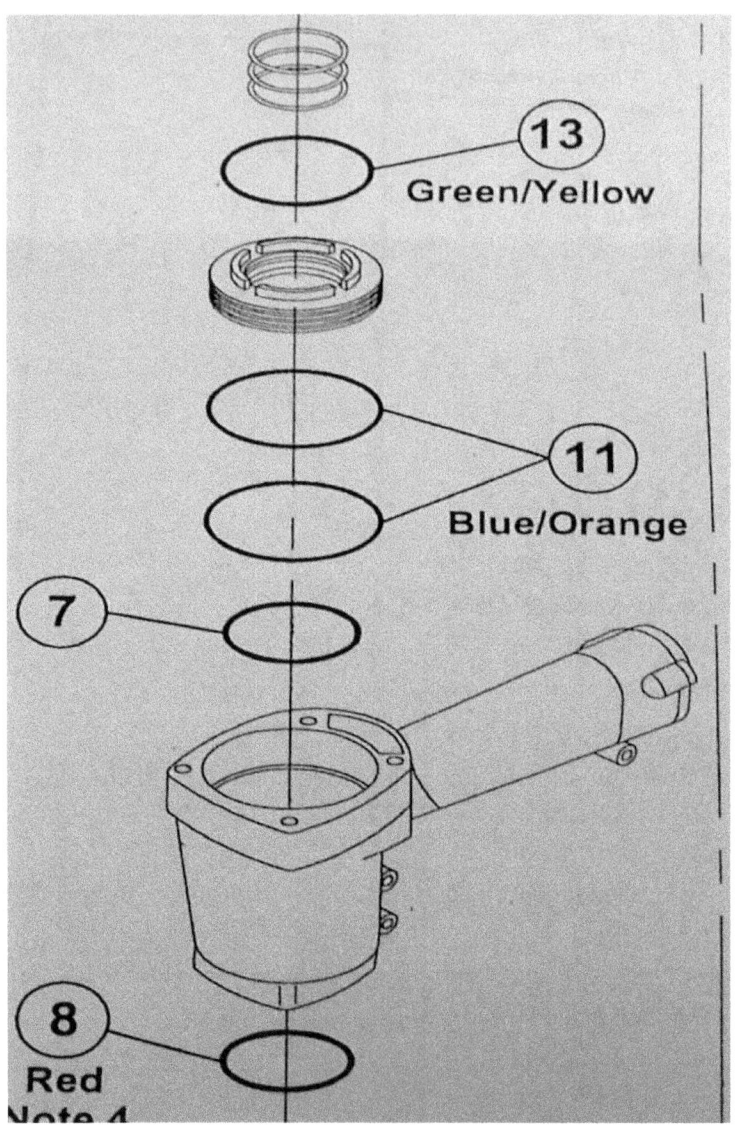

As you can see by the above schematics how the parts and o rings line up to go into the tool. The schematics show what goes into the cap, the body and the trigger. There are a lot of variations on how the schematic is drawn and what information it will contain by brand.

If a schematic exists, print it out first thing. Look at the exploded view and then go to the parts list. If you see the words, "obsolete", or "no longer available", you may not be able to repair your tool. This can save a lot of time tearing your tool down and then going to the schematic and parts list. It always seems to me that the part I need is the one that is obsolete.

This is where your local repair shop will come in handy. Some nail gun parts can be interchangeable. If someone has the experience repairing different brands and model guns they may be able to substitute a part and make it work. I have even taken bottom bumpers and sanded a bevel on them to fit a gun, and it worked.

I can't tell you how important the schematic is. It helps taking the tool apart and putting it back together. You really can't order parts without the part numbers that are on the schematic. When researching tools, check all the parts on the schematic. Schematics have a lot of detail, but sometimes the detail is not great enough. Pay attention when taking your tool apart, it will help greatly. You can also put the parts in order as they come out. You can use a permanent marker to number parts. Make notes if you think it will help.

Chapter 2
Preventative Maintenance Tips

Electrical Powered Air Compressor, Oil Lubricated

These are things you need to check to keep your compressor running with less down time. A little effort and money will save you a lot in the long run.

With an electric motor to run your compressor you need to have the proper cord on the motor. The cord that it comes with is the appropriate size for its length, but if you change it, replace it in kind. Always put the compressor next to the outlet and run the air hose to the work area. If you use an extension cord it must be the correct size or the motor will burn up.

An explanation of how to choose an extension cord follows. The extension cord is sized by the amps of the tool. This is the flow of the current through the wire. If the tool can't get enough current it will burn up. I am listing some average amps per tool and then the cord lengths.

1. Circular saw 12-15 amps
2. Standard power drill 3-7 amps
3. Hedge trimmer 2-3 amp
4. Weed trimmer 2-4 amps
5. Chain saw 7-12 amps
6. Leaf blower 6-12 amps
7. Bug lamp 1-2 amps
8. Lawn mower 6-12 amps
9. Table saw 15-20 amps
10. Sawzall 6-8 amps
11. Router 4-6 amps

Remember that as the wire gets thicker the more electricity, or amperage it can handle. Just because a cord is thick doesn't mean the wire inside it is. Most cords are written on with the gauge wire or imprinted into the cord. Here is a list of cord lengths and amperage it can handle.

1. 16 gauge cord from 0-100 ft long will handle up to 10 amps.
 2. 14 gauge cord from 0-50 ft long will handle up to 15 amps.
  3. 12 gauge cord from 50-100 ft long will handle up to 15 amps.
  4. 10 gauge cord from 50-100 ft long will handle up to 20 amps.

Some compressor manufactures recommend an 8 gauge wire for over fifty feet. This may be because of a peak pull of electricity or amperage on start up. If you are in doubt, go to the larger gauge to be safe. You can see by the chart, if you went with the 8 gauge you should be safe under any condition. This can be expensive, but you won't destroy your tool.

If you want to run multiple tools on one cord you will have to add all the maximum amps together. If you do that, you can see there are few tools that can be run together at the same time. A homeowner can do this because he will only be able to run one tool at a time. In the construction world where there are multiple people, this will not work. Extension cords are expensive now, but so are the tools that can be burnt up using the wrong extension cord.

If you want to be safe with your compressor, use the air hose to go to the work area, it is cheaper than an extension cords.

There is a belt from the motor shaft pulley to the compressor shaft pulley. These pulley set screws should be locked down with a dab of locktite. You should make sure the belt is straight. If not the belt may come off, slip, or wear out. The belt should be tight, but not too tight. You should be able to push the belt down in the center about a quarter of an inch. If the pulley gets loose on the shaft it will waller out the pulley center to where it will need to be replaced.

The compressor, or air pump, has an oil level in the bottom of the case. Some have a sight glass with a max fill line or dot. Some will have a dip stick with a max and minimum fill level. If it doesn't, check the depth of oil from the add oil opening to the bottom with a stick. If it is an inch or just over, that should be sufficient. If it is several inches you will have to find out exactly how much oil to use. It doesn't take that much oil to splash the internals for lubrication. This is a special oil sold at tool shops or hardware stores. If you have too much oil in the case it will slow down the movement of the pump which will put the motor in a strain and may trip the breaker. So, more oil isn't better.

Another area not taken seriously is the filter on the suction of the compressor. If this is plugged, you won't be able to get a good flow of air into your tank. If the filter has a hole in it you will suck up dirt and dust into your compressor, tank, and into the air tool.

The air tank, or tanks are also important, especially in high humidity areas. When the air cools in the tank the moisture falls out. Your tank should be drained daily. When water is carried to your tool and mixes with the oil, it creates an emulsion. This emulsion lowers your lubrication's ability to do its job. Less lubrication means more wear. If you think that the rubber o rings will wear and not the metal, think again, they both wear. The moisture can also cause carbon steel parts to rust, possibly locking parts up where it is unable to move.

Gasoline Engine Powered Air Compressor

The pulleys and air compressor are similar with both motor and engine operated compressors. The difference here is the gasoline engine.

The major area is the oil level and changing the oil on a regular basis in the engine. Use the engine oil recommended by the manufacturer. Some engines have a low level shut off on the case. Keeping the correct level in both engine and pump is important.

Oil less compressors may not be cost effective to repair. When you suspect what is wrong with the compressor, go to a repair schematic on the Internet and see what is offered. If there are no schematics of the motor or compressor, it may only be sold as a unit or not at all. Most of the small compact units fall into this category. Oil less compressors will be cheaper to buy than to repair. You can compare the prices with a schematic and parts list to see the actual cost.

Checking your air filter regularly is important for both compressor and engine. If you are in a dusty area it should be changed more often. It would be a good idea to have a spare filter on hand.

The fuel is important, if the engine sits up unused the gas will go bad and clog the small ports in the carburetor. Some sort of fuel stabilizer should be used in the engine gas tank if it will be down for a while. The problems I see are caused by fuel sitting up more so than anything else. If you have a stabilizer in the fuel and you shut off the gas and run the carburetor empty, that will help.

Ethanol is another problem. If you use non ethanol fuel that would be better. If you have to use it, use an additive for ethanol. Ethanol is alcohol and will dry out or melt certain rubber parts in your engine. Draining the tank and running the engine until it stops is the best way to avoid fuel problems. Fuel problem repairs usually will run about one hundred dollars.

Electric Powered Air Compressor, Oil Less Compressor

All of the above tips still apply here. The only difference is that this compressor does not have oil to lubricate the internals. Oil less compressors are made to run a set number of hours and some part of it will fail. Usually when the motor wears out, it is not cost effective to repair. Most of these compressors are sold as a unit of motor and compressor. If the compressor end fails there could be some individual parts offered and that could be cost effective if you do the work.

Look at your schematic to know what parts you have in the compressor and how they go together. Does the compressor have a belt, piston ring, breaker, or reset button? Where is the suction filter? Some companies will replace the piston o ring and some will replace the whole piston. This information is on the schematic.

The more work you do yourself the less money you will spend. You can locate part numbers on your schematic and search for the best prices.

Generators

The generator can have problems when running long hours in emergencies. More problems arise when they sit up and the gasoline goes bad. When gasoline goes bad, it turns to varnish inside of the carburetor. If the gas will be left in the tank you should add a fuel stabilizer. After adding the stabilizer start the engine and run it for a few minutes and block off the fuel. Let the fuel burn out so the carburetor is empty. Ethanol is a problem here also. Try to stay away from it if you can. If you can drain the fuel tank, that will also help.

Some of these engines have a low oil level shut down. So if your engine oil gets low the engine will shut off and will not start until you bring the oil level back up. When running a lot, the oil should be checked and changed more often.

The air filter also should be checked more often as it runs longer. A dirty air filter can stop an engine from getting the air it needs to run. The engine may shut down or you may not be able to start it, simply because the air filter is too dirty.

You should always read and keep the documentation that you receive with the generator. This will tell you what grade oil to use and all the things you need to know to crank and run it. You will need the model number to be able to order parts or look up schematics.

Do not run a generator inside, it needs to be outside or the exhaust can make you sick or kill you.

Since this is a large electrical generating piece of equipment, keep it away from any water. You could get shocked or destroy the electrical end of the generator.

Pneumatic Nail guns

There are a few small things that will extend the life of your nail gun. One is keeping the inside of your air hose clean. If your air hose has been drug through the sand, you are going to get sand in your gun. This sand will wear out the metal as well as the o ring seals. The hard parts of a nail gun will run your repair cost up dramatically whether you do the work or not.

The next item will keep your gun working and working well. This is lubrication, pneumatic air tool oil. Never ever use WD40 inside an air tool. You can use it on the outside to clean it up, but you must blow it off dry. If you don't, the dust and grime will stick to it and you will have one dirty gun. Using the wrong oil can attack the o rings and make them swell, dry out or get sticky.

These tips will work with any air operated gun. Take a look at your gun's schematic to see how parts are lined up. When you access the schematic on line it will give a parts list and have the parts numbered on the schematic so you will know what it is called. You can use these part numbers to order from any vendor selling that part.

Some pneumatic guns are oil less. If you put oil in them it will destroy the rubber diaphragm and the gun will not work. Read your gun's instructions. Most manufacturers have an oil less gun or even several models. Other brands may not be so adversely affected when oil is mistakenly put in it. It may just gum it up.

For a gun that uses oil I would recommend ten drops for a work day. If you use too much it will get all over your hands and the tool. Too little oil and it may lock up or cause wear to the internals of the gun.

Moisture coming from your compressor will mix with the oil and create an emulsion. This breaks down the oil and your tool is no longer getting the lubrication it needs. Emulsion will be white and frothy.

Gas Operated Nail Guns

The gas operated guns are a little different to maintain. The most common problem over time is that they get dirty inside and the spark plug fouls. In most cases a good cleaning will get them going. If a gun has some age on it and breaks down, you will have to replace some parts. They have o rings, an electrical circuit, an exhaust muffler, fan, and moving cylinder. I can tell you that whatever you have to pay a trained repair mechanic, it is worth it. These guns are complicated. They have the tools to do the work and it will save you down time and money in the long run.

Dropping the gun is the most common problem. Having a good battery and back up is also necessary to keep working. Batteries do go bad and will need to be replaced. Your charger may also go bad. You can use what you learned about the VOM to check the cord and if it is putting out the correct voltage to charge the battery.

When purchasing the gas canister you need to check the date. If the date is over a year old, don't buy it. These canisters will go bad over time.

Drills

One part of a drill people overlook is the chuck. There are a lot of moving parts in the chuck whether it is keyed or key less. Spray some lubricant inside the chuck and work it open and closed. After this exercise, blow it out with air. Chucks are made of different material. If it is carbon steel you have to keep it away from water or moisture.

If you should open up the case of your drill to check it out, be careful. Some of these drill parts will only fit one way, or the case will not close. Be sure to pull the brushes out first. If you don't, when you pull on the armature, it may damage the brushes. When you get ready to put the armature in the brushes cannot be in, they will stop the armature from seating. This will be true for any tool that uses brushes. They will be the first thing to remove and last to install.

A drill that operates on AC current will most likely have gears or a gear box that can be greased. DC drills also have gearboxes that can be lubricated but it is very difficult. The DC impact or drill will also have a transmission. When you take some of these DC drills apart, motor from transmission, the gears can fall out. Be very careful. If you lubricate, use a light grease or a heavy oil.

Hammer drills have a lot of moving parts that can be checked for proper lubrication. A light grease will work well in any drill gearbox. The end of the hammer drill that holds the bit needs to be lubricated, but blow the excess out. Too much oil will cause it to clog up. Not enough and it will wear out or seize up.

The extension cord will play a big part in your drill working properly and not burning up. Refer to the prior chapter on extension cords. If you replace the cord on the tool, replace it with the same or larger gauge cord.

Blowing the motor out with compressed air will keep your motor and case clean. Air needs to circulate through the motor to keep it cool. You can do this without having to open it up.

Blow through all openings to remove sawdust or other materials. When heat builds up in the motor, parts start to break down. This dirt could also cause a short in the windings of the field or armature. In most cases, if the armature and field burn up, it isn't cost effective to repair it unless you do it yourself. You can determine this by looking at your schematic and pricing the parts.

Saws

There are a lot of different saws now days. There are circular saws AC, worm driven saws AC and battery operated. Then there are variations of each one of these. Most of the worm driven saws I have seen do not have the proper oil level in the gearbox. There is a plug that should be removed to check the oil level in the gearbox. The saw should be sitting up level, like it is cutting a board. The oil level should be at the bottom of the plug hole. There is a special oil sold for these gearboxes. In this instance, you can't put too much oil in the gearbox. Some brands of worm drive saws will have a closed gearbox and cannot have lubrication added. I don't like those because sometimes you can't stop a seep and then you have no oil and the gearbox goes out. If you see oil around the gearbox and sawdust stuck to it you should check the oil level regularly.

A circular saw will generally have a small gearbox that isn't easy to get to. The older the saw, the more important it is to check the gearbox.

Other type saws such as a jig saw, sawzall, or saber saw will need lubrication. When you open them up to check, take it slow. Most times a light grease is what you need to use because oil may leak out. Make sure not to overfill the gearbox with grease. A small amount will work well. As the gearbox warms up the grease will flow. If the grease is too thick it might not flow to lubricate the gears.

When using a battery operated saw you need a good battery. Most importantly you need a sharp blade. You don't have the power that an AC motor has, so the sharp blade will help do the best job possible.

Your AC power source is important. Your cord should be the correct size with no cuts in it. Your plug should also be in good condition to keep you safe.

Your DC power source should have a good battery and charger. You should check your battery and tool contact points for burnt spots or corrosion. Tight connections also help.

Sanders

There are a lot of different sanders out there. Very few will have a gearbox. Some may need a shot of oil here or there. Some bearings may need replacing at times because they are closed and cannot be lubricated. There may also be bushings that need lubrication.

Again, the extension cord will be important in the operation of your sander. You will check it the same way we did in earlier chapters. Use your schematic when doing any kind of maintenance.

With all the dust made by a sander, blowing out the motor is most important on a regular basis. Your motor fan can't cool the motor properly if the windings and case vents are plugged.

Take a look at the schematic and look at the parts. You can see ahead of time what you are looking at, or what you are looking for, and where it is. This schematic has electrical as well, so that you can see how the cord wires go to the switch and leave the switch for it's circuit.

Some sanders will have a belt of some sort to move the sandpaper. They can also be direct drive or have a gear. They can be complicated, so take your time when taking them apart.

Chapter 3

Repair Guide

In General Repair

Some of the checks performed will be the same for several tools. For instance, the electrical cord. The cord can have a broken wire inside or cuts in the insulation. Cuts can be taped up for safety. You will have to decide where to start your diagnoses from the problem you have. For instance, your compressor isn't filling the tank with air, but the motor or engine runs. You would start with checking out the compressor end, not the motor, or engine. When I work on something in the shop, I check it out end to end. You can use the information in this book in part or in full. I can't give specific information on every part of every tool because there are so many combinations. Therefore I will try to do it in a general way so that you can use this information on many different tools.

We will start by checking out the cord. You will need your volt ohm meter and what you learned about how to use it earlier. The black probe will be plugged into the common socket and the dial to ohms. Some volt ohm meters will make a beep if the circuit is complete. That will be the least amount of power pushed through the circuit. Go ahead and turn your selector to the highest ohm number and watch the needle or readout. This will insure you are sending enough power through the circuit to check it. Touching your two probes together will give you a reading and tell you the VOM is working.

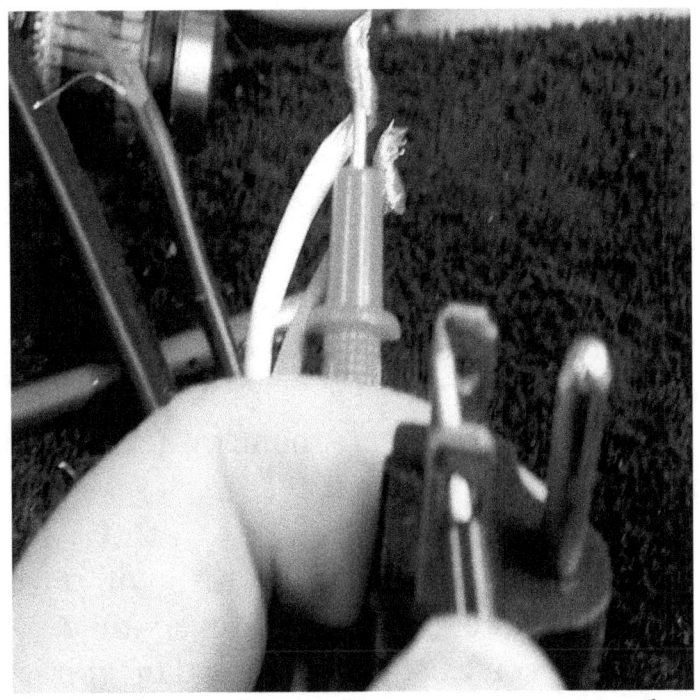

In the above picture you can see one probe on the white or black wire and the other on the plug prong. The black probe appears to be touching both prongs but it is not.

The cord can be connected to the motor, or disconnected to check it. Hold the plug close to the motor end of the cord. Take your black probe and hold it to one of the prongs on the plug. Now take the red probe and touch either the white or black wire at the switch. This is where the end of the cord actually is. If you get a reading, the wire is good. If you don't get a reading, touch the other wire. If both wires show a reading, then both wires are good. If one of the wires does not show a reading, it has a break in it and is bad. Now you know if the cord is good or bad. If you have a ground wire, check it the same way. Check for cuts or nicks in the cord where water could get in and shock you. Contractors or home builders have to have a cord with a molded plug. This is an Osha rule to help protect the user. A homeowner can have any kind of plug he pleases.

When checking other tool cords I will refer to this procedure. It is the same every time. Sometimes you can cut the bad end off if it is at the tool end. Just cut a little off and recheck the wires. If they both show good now you can reattach it to the tool. It may not matter to you if the cord is long or short. If the broke wire is at the plug, replace the cord.

Next, check the motors reset or breaker, if it has one, and reset it. It is not in the same place every time. It could be positioned close to the motor or not. Your schematic can help locating the breaker if it has one. You can check the reset with your volt ohm meter. When the reset is pushed there should be continuity from one side of the reset button to the other. If you don't have that closed circuit then the reset button is bad. A breaker will work the same way.

Some motors will be easier to check than others when looking at the field and armature. The field is the winding around the inside of the case. The armature is the shaft and coil that goes into the center of the field. The area the brushes ride on the armature is called the commutator.

If you know how something works it enables you to better diagnose a problem. I will explain how the simple motor works. This is the motor used in most small tools you will be working on.

We will start at the plug that goes into the wall socket. One side of the plug goes through the cord and enters the field on one side of the motor. A small motor has two coils in the field. These two coils are turned into electromagnets, one a south pole and the other a north pole. The wire then exits the field and goes to one of the brushes on the same side of the motor. The brush is touching one block on the commutator which is connected to one coil on the armature.

The current flows the opposite armature coil that is touching the opposite brush. Now the current flows from the brush to the opposite field coil, then out of the field coil. When it leaves the field coil the current passes through a switch and back into the socket.

The armature coils when magnetized are opposite poles to the field coil. As the armature tries to move itself to like poles the brushes move to another bar on the commutator. This now creates a new electromagnet on the armature that opposes the field coils poles. The armature is in a futile attempt to align its magnetic poles with the fields poles.

So you can check this circuit from the cord plug. You will have to pull the switch to the on position. If there is a break anywhere in that circuit you will not get a reading on your VOM. Now you have to check each part in that circuit separately. These areas would be the cord, brushes, armature, field, and switch.

When you check the field with a metal case motor, it can have an open wire, or a ground. This means a break in the wire or the wire is touching the case. Each field coil has two wires to check on the circuit. There is a wire going into the coil and the wire coming out of the coil, which is really the same wire. Using the VOM, set to the max ohms for this test. Hold the black probe to one wire and red probe to the other wire. If you get a reading, the circuit, or wire is good. If you don't get a reading, the field is bad. If all field circuits are good, the field is good. Each field can have two or more windings.

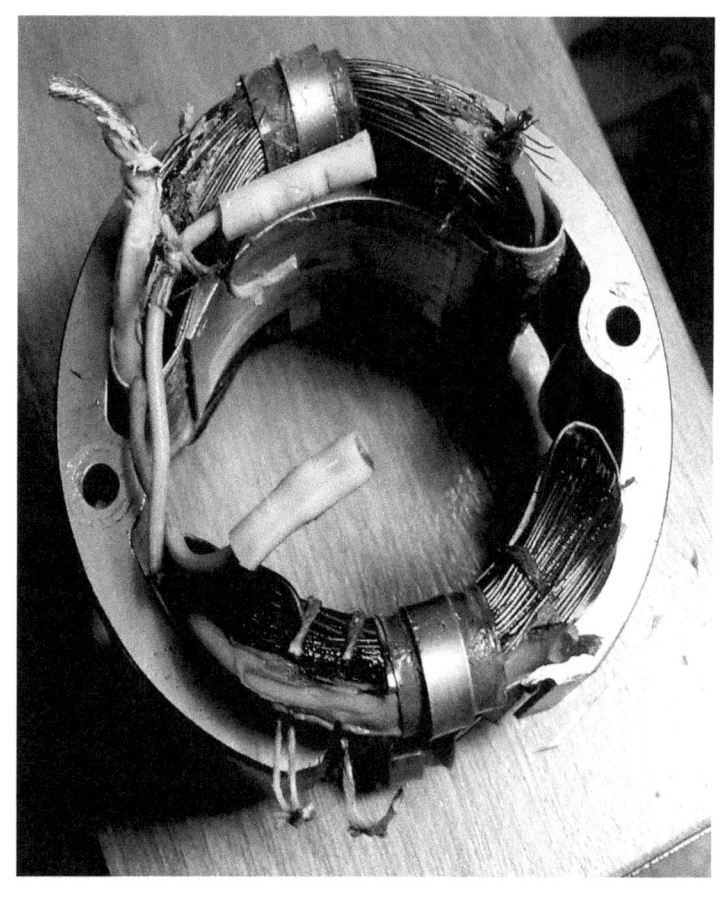

In the above picture you can see that this field over heated and melted the wire insulation. I will use this field to show how to check the continuity of each coil. In this case it is two.

In the above picture I am checking wires in pairs to see what is continuous and what is not. One pair would be a wire going into the field coil and one coming out going to the brush on the left side. On the right side the would be the same. A wire coming out of the field coil and the wire going to the brush

Now, how to check for a ground on the field. Use your VOM selector at ohm. Connect the black probe to a wire of one field coil and the red probe to the case. If you get a reading, you have a short. This means the wire is touching the case. This circuit will not let the motor work. Check each field coil the same way. This check can also work in a generator field gone to ground.

You can now check the armature. Using the VOM for ohms again. Since there are no wires sticking out of the armature as the field does you will have to check the armature through the commutator. The wires that are wound around the armature end at the commutator. This is the area that the brushes ride on.

The above picture shows the probes on the commutator. Keep the red probe where it is and move the black probe down to the next wire or flat surface. Do this all the way around to check each circuit.

One thing I forgot to mention was that when you are looking at a field and armature, look for burnt spots. This is a sign of a short or overheating. Loose connections also cause sparks and heat. If there is metal missing, the armature will eat up the brushes and should be replaced. Any sign of extreme heat is not good. If the motor still works it will continue to deteriorate and will eventually stop working.

Now, hold the armature in your left hand with the black probe of the VOM on one commutator section. With your right hand and the red probe, touch each remaining commutator section. If one section doesn't have a continuity reading, the circuit is bad and the armature is bad.

Your bearings also have to be checked. Bad bearings can cause higher amp pull on the motor and make a noise in the motor. If the bearing is tight and spins freely, it is good. Sometimes you can tell without a doubt that the bearing is bad. If the bearing feels rough, but turns freely, replace it. It will go bad.

To remove a bearing you need a bearing puller and an arbor press. The puller will be spread apart to slip over the bearing and placed on the bearings back side. Use the bolts to close the bearing puller. If the bearing doesn't come off by closing the puller, you will need to use the arbor press to push the shaft through the bearing.

When using the arbor press to push the bearing onto or off the shaft, you need to protect the bearing. The bearing can be damaged if you are not careful. Use a large washer or socket that will exert equal pressure all the way around the bearing. The race is the area where the balls ride and is the easiest to damage.

When you install a motor on your air compressor, the belt has to be lined up straight. The motor pulley and the air compressor pulley should be in line, or the belt will wear out, or come off. The belt has to be tight, but not too tight. In the middle of the belt, take your finger and push down about a quarter of an inch. Too tight can stretch the belt and too loose it will lose traction.

Air Compressor Repair

Picture #1

Now we will explore the compressor end of the air compressor (#10 in above picture). Check the compressor belt pulley that is on the shaft and tight. Check the bearing that holds the shaft in the case housing for wear or looseness. If it wiggles, the bearing is bad. If anything is worn, it may have worn out other parts as well. Look at your schematic and your compressor. Check the parts availability and cost. Determine if you think you can make the repair and what the cost would be. Then get a price on a new or rebuilt compressor to see which is most cost effective.

Is the oil level OK? Does the air filter need changing? If the compressor is not pumping air, there can be one or several problems. The piston has rings (#6 in "Piston Upstroke" picture below)

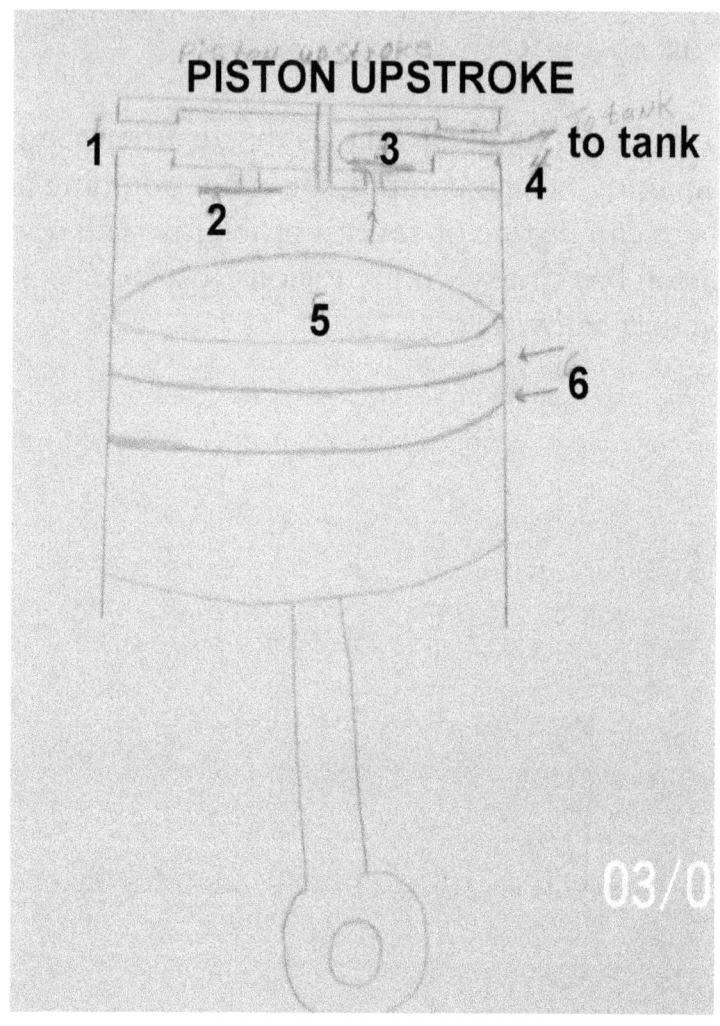

PISTON UPSTROKE

1   3   to tank

4

2

5

6

03/0

that seal the piston in the sleeve. There can be one to multiple rings on the piston. It can also be a synthetic material that goes between the piston and cylinder wall. If they are broken or worn out the pressure will leak by the piston. Check to see if there is a positive pressure in the compressor crank case or if the oil is frothy. If it is, the rings can be leaking air by into the crank case. There are reed valves (#2 & #3 in above picture) in the top of the compressor that open and close to suck air in and then have air pumped out to the tank. One of those reeds can be broken. The above picture shows the reed valves in the upstroke position. The complete cycle takes place when the piston goes down, opening the suction reed valve and closing the discharge reed valve #3. The air flows through the suction filter #1 into the suction chamber, through the reed valve #2 and into the piston cylinder. When the piston moves upward, the air flow closes #2 suction reed valve and opens the discharge reed valve #3, and the air flows through the discharge line into the tank.

You can purchase parts locally or on the internet. My experience with compressor repair is that if a shop does it, you can buy a new or rebuilt compressor pump as cheap as having one fixed. But if you are going to do the work yourself, you may come out ahead.

When you start the work on the compressor end you will need a schematic. This will help you take it apart and put it together. It will also help you order the parts you may need.

To get to the piston and reed valves you must disconnect the discharge piping and remove the top of the compressor cylinder. Be careful with the gaskets. If they are torn or questionable, replace them. When taking the compressor apart, go slow. Make notes if you think you will have trouble remembering how it goes back together. Use a permanent marker to mark pieces front, back, top, or bottom. The schematic may not have enough detail to show some positioning.

Inspect all parts in the top of the compressor including the gaskets for cuts or breaks. If there is any question, replace it.

The compressor also has a check valve and an unloader valve. The check valve keeps the air flow going in one direction and gives the piston the ability to increase the tank pressure to its maximum. The unloader valve will come in many shapes and sizes. It can also be called a pilot valve. When you see a small line coming off the compressor head, #9 (picture #1 above) and going to a brass object, that is the unloader valve. This valve takes the pressure off the head or piston of the compressor so it can start. With pressure on the piston, it is very hard to impossible to start. Look at your schematic to find these valves. This valve can be broken, plugged or leaking.

In picture #1 above you see the discharge line #8 going to a manifold. On that manifold sits a pressure switch #6 and pressure gauge #7. Both of these items are looking at the pressure in the tank. This pressure gauges tells you tank pressure.

There will also be a pop off valve (behind #7 picture #1 above) on or near this manifold. This is a preset setting to pop off and let excess pressure escape so your tank and fittings will not fail. Your compressor will have a maximum operating pressure and that is what this valve will be set at. It is not adjustable. They can be replaced, but it should be the same pressure setting.

Pressure Switch

There is also a pressure switch. This switch is operated by the tank pressure to start and stop

the compressor motor. You will see the electrical cord goes into the pressure switch box #3 (upper pic) and the two wires are screwed to a contact marked line. On the other side of the switch (lower picture) #4, a cord goes to the motor marked motor. This switch is what cuts off the power to stop the compressor from pumping too much air into the tank.

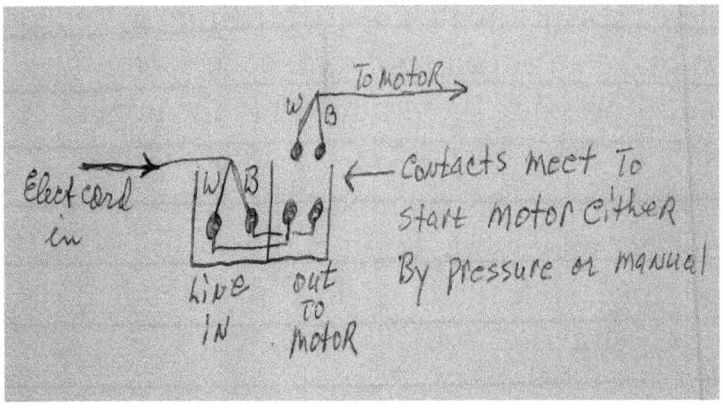

This is a simplified drawing of a pressure switch

There is also an on off switch that can be configured a lot of different ways (upper pic, #4 would be a push pull button). There is an adjustment that allows the motor to shut off at a designated pressure (lower pic) #3. The pressure switch is positioned so it can see the tank pressure. When the set pressure is reached, the electrical circuit is opened and stops the motor. The spring you see in the picture below is used for the tank pressure to overcome and trip the switch. This switch is broken or damaged by people more than any other part on a compressor because they think they can fix it.

The pressure switch is not repairable. You can make an adjustment to raise or lower the shutoff pressure by turning #5 in the upper picture. There are a lot of different kinds of pressure switches. Some that are just a little different may work, but it may have to be that exact one. It depends on how it is mounted, lines going to it, and other variables.

The yellow piece (#2) in the above picture is the pop off valve. It is made into the pressure switch. So you can see, they can be mounted anywhere it can see tank pressure.

Pressure Switch

This picture above shows #1 where the contacts are housed for the switch. #2 is the adjustment to increase or decrease the pressure setting shutoff. #3 is the lever that operates the contact points on the other end starting and stopping the motor. #4 is the wire coming out of the switch going to the motor.

Motor

The motor will have one or two capacitors, depending on the motor. One will be a run and the other a start capacitor. These can go bad and the motor will not start or it won't run. You will need to take them off and test them. Use the methods discussed in previous chapters. They may have burnt areas or melted plastic to help tell you that it is bad. If you are in doubt, have it checked at a repair shop.

Check to see if the motor has a reset or breaker. One of these could be tripped or broken.

If there is still a problem with the motor you will have to check the field and armature as we discussed in a previous chapter.

There could be a broken wire, bad field or bad armature. Start your check with the cord going to the motor. Then the capacitor and reset button. Then check the field for a ground or a broken wire. The armature should also be checked for a broken wire.

Discharge Manifold

On the piping leaving the tank is a pressure regulator. This is the way you raise or lower the pressure that will go to your air tool. It may take you several times adjusting it to get it on the correct pressure. If the pressure won't regulate, replace it. The regulator is not repairable.

After the regulator, in the piping, there will be a cutoff valve. Then you will have quick connects for the tools to be connected to. The number of connectors depends on the volume of air flow at a set pressure your compressor can provide.

When doing any piping work on the air compressor air lines, use pipe tape when screwing parts together to avoid leaks. When the quick disconnect fittings get loose or start leaking, replace them. On the inside of these quick connects are little balls that hold on to the male fitting. The more of these balls you have the better it holds the male fitting and the less likely it is to leak. So a fitting with six of these little balls would be better than one with three. The cost also goes up with the number of balls.

There is also an oring down inside. If you oil the connector do it with tool oil. If the connector leaks, it could be the oring has broken.

Engine Operated Air Compressor

Picture #1

The engine operated air compressor has the same general compressor, but there will be some different valves used.

There is also a safety valve, #2 (pic #2), that is preset to pop off to keep from over pressuring the tank. If this is leaking it will have to be replaced. Make sure it is the correct safety valve for your compressor. Some times the pressure setting is stamped into the metal on the valve. You need to know your compressor tank's max pressure to make sure your safety valve will protect your tank.

There is an unloader or pilot valve, #1 (pic #2) on your compressor. The check valve, in the discharge line at the tank, is to keep the pressure in the tank. The unloader valve does three things. The first is to idle the engine down, it takes the pressure off the cylinder head, and stops the air flow to the tank. #9 (pic #2) is a manual unloader. If it is pointing up it will unload the compressor so it doesn't pump more air into the tank. It has to be up when starting the engine or it won't start because of the cylinder head pressure on the pistons. When you move it to the side and the stem goes down, that starts the compressor pumping into the tank again. If either of these are not working, the compressor will not restart or pressure up the tank. The line going from the unloader valve to the tank

Picture #2

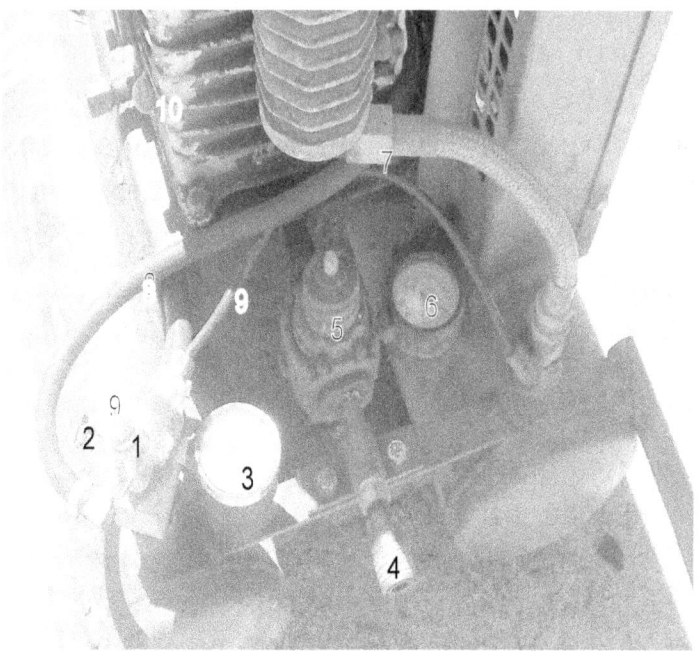

discharge allows air to flow out of the unloader valve so the tank doesn't pressure up.

There is a spring in the unloader valve like the pressure switch that lets the compressor unload at a certain pressure. #8 (pic #2) is the line going from the unloader to the engine idle mechanism. When the unloader is unloading the compressor it manipulates the idle mechanism to idle the engine down a bit. #3 (pic #3 below) shows the idle mechanism and how it is connected to the engine idle. As the stem moves in and out it moves the engine to idle or run for compressing air.

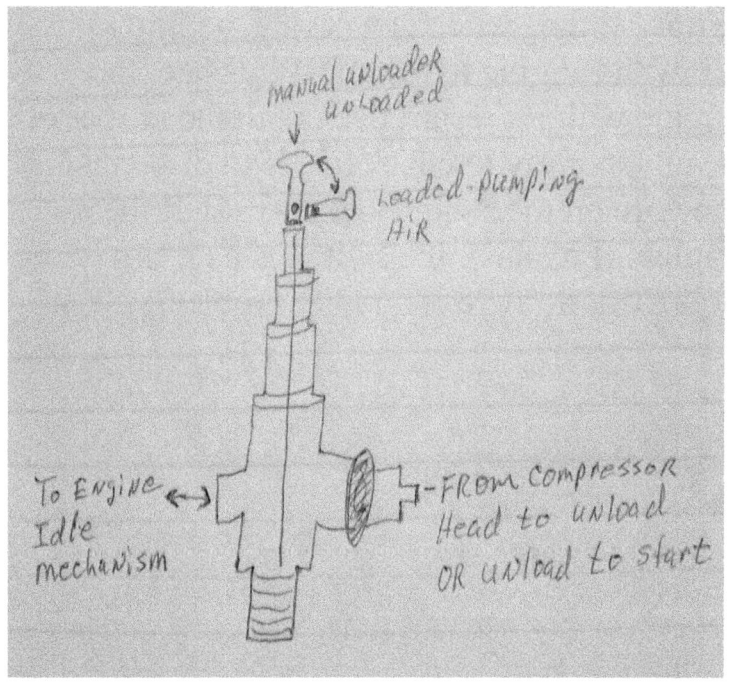

The above picture is of a pilot or unloader valve. The most important part of this valve to you is the top toggle lever. You can see straight up it unloads the compressor and when on its side it loads the compressor. Load meaning pumping air to the tank to be used. This toggle is manipulated by you manually to start the compressor and then start pumping air.

In picture # 3 below, you can also see the engine oil level cut out switch (#1). If your engine oil level drops below a set level this switch will kill the engine and it will not restart. If the switch is working, adding oil will allow the engine to restart. If it is bad, the engine will not restart even with the proper oil level. One way to check this is to disconnect the wire on the switch and try to start the engine. If it starts, the switch is bad. If it doesn't start, there is another problem.

Picture #3

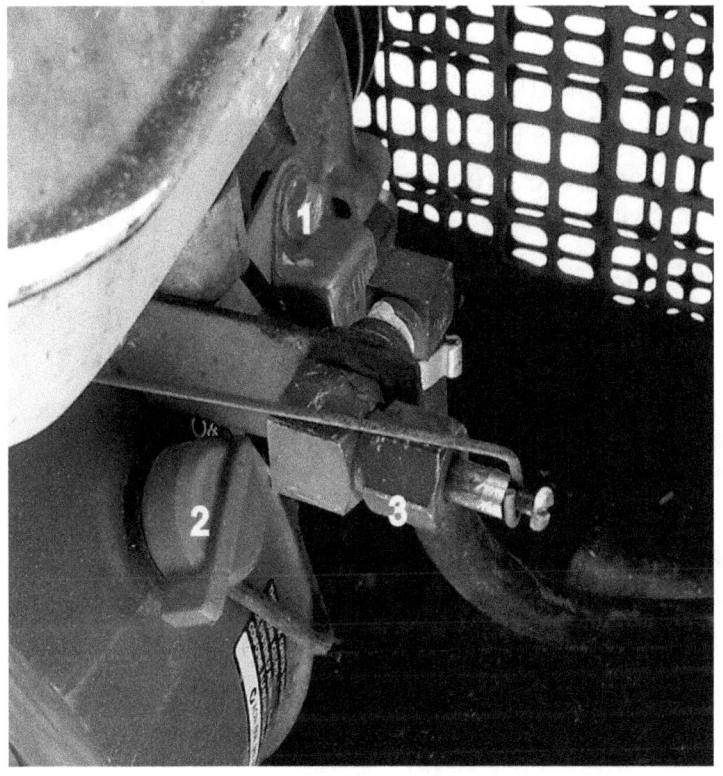

#2 in picture #3 is the on/off switch to start and stop the engine. This will be in the on position in order to start the engine and the off position to shut it down.

In picture #2 above, #2 is the safety or pop off valve. #3 is the pressure gauge showing the tank pressure. All of these items are looking at the tank pressure to do a particular job.

#4 in picture #2 is a quick connect for your air and #5 is the pressure regulator. This regulates the pressure you will use in your air tool. If this regulator goes bad, replace it, it is not repairable. #6 is the regulated pressure reading of the air going to the air tool.

So if your compressor has a max of 150 pounds, the tank pressure gauge should not be over 150 pounds. If your tool has a min/max pressure of 90/120 pounds your regulated pressure should be between 90 and 120 pounds. If your tool will work at 90 pounds, that will be less stress on the tool. As for the compressor, it will allow the compressor to catch up faster.

#10 in picture #3 is the crank case where the oil bath is located. The crank shaft that the pistons ride on will travel through this oil and splash it on all the internal moving parts. Good clean fluid oil will get to all of those parts. Too much oil puts a strain on the motor or engine and could cause the oil to leak out through air vents.

The tank or tanks will have drain valves on the bottom and will be in all kinds of configurations, but they all do the same job. The difference will be the size of the opening they screw into. Draining the tanks is just as important as any other part of the compressor or engine. If the bleeder valves leak, you will not have water in the tank, but your compressor will run more to keep up.

In picture #4 you are looking at a side view of the engine. #1 is the muffler, #2 is the air filter housing which also houses the fuel shut off and choke. #3 is the valve cover. This is where you would adjust the valves. #4 is the spark plug. Make sure you have the correct plug or it may punch a hole in the piston. #5 is part of the idle and run mechanism. This is where the unloading mechanism is attached to idle the engine down and up.

When working on engines and air compressors or any tool, be careful when tightening screws and bolts. The plastic, cast iron, and cast aluminum will crack or break if tightened too much. If you are concerned about the screw or bolt coming loose, use locktite or lock washers.

Picture #4

If you know how something works it will help you to analyze what the problem is with your tool. This applies to any tool. Understanding how all these different pieces work to make an electric or engine operated air compressor work will help fix the problem.

Here are some problems and the solutions:

The engine won't stay running.

Check: 1. The oil level switch or the oil level. 2. The air filter. 3. The unloader valve.

The oil level could be low or the level switch could be bad. The air filter could be so plugged that air will not flow through it. Take the filter off and see if it starts. The unloader valve may not be taking the pressure off the compressor head. The engine can't turn fast enough to start or if it does it will shut down quickly.

The compressor won't pressure the tank up.

Check: 1. The pilot valve. 2. The check valve. 3. The reed valves in the head. 4. The gaskets in the head.

The pilot or unloader valve could be stuck or broken. The check valve could be stuck open. One of the reed valves could be broken. One or more of the gaskets could be broken or dried out leaking either internally or externally.

Think of the air flow as it is going through all these parts. What could be stopping or diverting this flow. What part of an engine could easily keep it from starting. If the simple stuff doesn't work out then move on to the hard stuff, like compression and valve adjustments. Take one thing at the time and try to eliminate things until you come across the bad actor. This can apply to electricity flow, oil flow, nail movement, and more. Also, get as much information as you can get to aid you in analyzing the problem.

Generators

Some generators have brushes. If your generator has brushes, that would be a good place to start when doing maintenance or repair. How would you know if it had brushes? Look in your manual and check a schematic. If it doesn't have brushes, it will have some way to get the current from the stator to the rotor to magnetize it. There are so many different kinds, makes, and models. I will talk about repairs in general. You will need your owner's manual and a schematic of how the generator is put together to proceed. This will also give you a parts list to check parts prices with.

Sometimes the problem is easy to see. A loose wire, a burnt area, or a swollen plastic box. You will need your schematic to identify the part to order.

The multimeter or volt ohm meter is what you need to check the electrical parts. It should be a good VOM so you don't get any false readings. The electrical system is a circuit. If a diode is burnt, it will have an open circuit. The same goes for capacitors and voltage regulators. So when you check these parts you are looking for continuity or the lack of it. If the problem is that the generator is not producing power, get your multimeter out. Check all the wiring behind the electrical panel. Check the breaker to see if it is closed. When you touch each side of the breaker where the wires are attached, you should get a signal. If not, press or reset the breaker and test it again. If you still don't get a signal, the breaker or reset is bad.

Picture #1

While the generator is running, use the VOM to check the outlets for AC output. There is a difference if there is no current or a little current. Some current shows exciting the generator may start it producing electricity. No voltage could be a ground or broken circuit somewhere. Never plug an electric tool in to the generator until you have set the voltage between 110v and 120v. The tool could burn up due to lack of or excessive voltage.

A generator that uses brushes to supply the power should be checked to see if they are worn out or stuck. You will know if your generator has brushes by looking at your owners manual and parts schematic. Locate where the brushes are and check them. They should be clean and moving up and down in the brush holder freely. The area where the brush makes connection to the armature (rotor) should be clean. The brushes should have some length to them to be able to work. If they are too short, they may not make good contact, and need replacing. There is a spring that pushes the brush against the commutator that could be broken.

After checking all outlets, breakers, fuses and any other components dealing with the distribution of power, you can check the rotor (armature). When you pull the cover off the generator end you will have access to check the rotor. You will see two wires going to the brushes and the two rings on the rotor that they contact. If you use your VOM and touch the wires where they connect to the brushes (red to red and black to black) this will tell you if the rotor and brushes are good. Meaning the circuit is complete. The second test for the rotor is the ground check. Touch your VOM red probe to the red wire on the brush holder. Then touch the VOM black probe to the generator case. This tells you if the rotor is grounded. If it is grounded it will not generate electricity.

Now you need to check the stator or field. To do this you need to disconnect the power cable coming from the generator end to the distribution panel. Look and see what color the ground wire is and then locate it on the cable end coming from the generator. Looking at the plug end with the ground wire there will be an odd number of wire connectors. Have your VOM set to resistance or continuity, touch one probe to the ground wire at the plug end and the other probe to the generator case. If you get a reading the ground is good. If you don't, the ground is broken somewhere.

Now to check the coils that make up the stator or field. Place one probe to one connector in the plug end and start checking the other connectors with the other probe. When you get a reading on your VOM that means that circuit is complete. There can be two to three sets of wires making up the stator. You are searching for these two wire combinations. If all combinations show continuity then your stator is good. If one combination doesn't have a reading, that circuit is bad and the generator will not make electricity.

Now to check the stator for a ground. Place one VOM probe to one of the connectors in the plug end and the other probe to the generator case. Do this for each connector in the plug. If you get a reading then that circuit is grounded. This will not allow the generator to make electricity.

If you get good readings on each circuit for the rotor and stator you will now move to the capacitor or bridge rectifier. This is a twelve volt charge to make the electromagnet stronger. If the capacitor is bad it will not supply the voltage to the rotor to produce higher magnetism. If the bridge rectifier is bad you get the same result, no electricity.

To tell if the rectifier or capacitor is bad you will need to excite the generator. If you have a rectifier on your generator you need a small amount of magnetism in the rotor when the generator starts to make it produce electricity. If the generator has set up a long time it may not produce electricity when started. Nothing is bad, it just needs to be excited to get that magnetism back.

To excite the generator you need a 12 volt battery and two wires with alligator clips on each end. Plug your VOM probes into one of the outlets on the panel to observe voltage output. Start the generator, now connect the 12 volt battery red wire to the red wire on the generator brush holder. Connect the black wire to the negative side of the battery to the black wire on the brush holder. Watch your VOM reading. If you show 110 volts or close to it your exciting has worked. Disconnect the wires from the battery. Just to make sure, shut down the generator and restart it. If everything is OK the generator will still make electricity when restarted. If it doesn't, the rectifier or capacitor is bad.

As you can see, understanding how to use the VOM is important in order to assess if certain parts of a tool is working.

The last thing to do with your generator is to set the voltage. I prefer to set it at 115v, between 110v and 120v. The engine speed is what makes more or less voltage. So whatever you have controlling your engine speed will need to be adjusted. Use your VOM to observe the output while you adjust the engine speed.

The engine repair and maintenance will be the same as in previous chapters. Bad gas with no stabilizer, not changing the oil, not cleaning or replacing the air filter will cause problems that can be expensive.

In picture #1 I have numbered parts of the generator for their identification. #1 is the on/off switch to start and stop the engine. #2 is the low oil level cutoff switch to protect the engine. #5 is the engine governor that increases or decreases the engine speed. #6 is the back side of the control panel. This is where the breaker, sockets and anything else for electricity is located.

Picture #2

Picture number #2 above shows the engine part of the generator. #1 is the air filter and access to the carburetor. To take the carburetor off you will have to enter this area. #2 is the choke on the carburetor and you can see the linkage connected to #5 the engine governor. These work together to increase or decrease engine speed. #4 is another look at the on/off switch for the engine.

This engine works like a lawnmower engine. When you put a load on it the governor speeds it up. With some generators the more electrical consumption the more the engine speeds up to handle that load.

Picture #3

Showing some additional engine parts in Picture #3 above. #1 is the muffler. Make sure it is not plugged up. If it is, it will not start, it has to have an exhaust. Some may have a spark arrestor screen that can plug with carbon. #2 is the valve cover. This is where you go to set the valves. You have to get the information on valve clearance from your manual or the engine manufacturer's web site. #3 is the bowl of the carburetor. This is where a float rises to close off fuel to the carburetor or add fuel when the float drops down. You can also check here for water in case you are having problems with the engine starting or running. #4 is the fuel line coming from the fuel tank to the carburetor. Check to see if this line has a shutoff valve at the tank. This can be left closed and cause the engine not to start. It can also be closed to run all the fuel out of the carburetor. #5 is another look at the air filter compartment. You can see how it fits in front of the carburetor. So to take the carburetor off you would first have to take the air filter compartment off first.

Picture #4

Picture #4 shows the control panel for your extension cords. #1 is the generator where the AC voltage is made and sent to the panel. #2 is the reset or breakers for each outlet. #3 is the 220v outlet and #4 is the two sets of 110v outlets. Some panels may have more of less on them, this depends on its size and manufacturer.

Nail, Brad and Staple Guns

All of these type guns work on the same principles. The nose will be different to handle the type of fastener it will shoot.

Here are the basics. When the gun is connected to an air compressor, it uses that pressure and air flow to operate it. The air enters the handle and will flow through the trigger to the bottom of the body forcing the piston driver up. In this position it is ready to drive the fastener into the wood. When pulling the trigger the air will then enter the cylinder from the top and push the piston down. There is a driver shaft attached to the bottom of the piston that hits the top of the nail as it is fed into the nose from the magazine. As the piston reaches the bottom of this stroke, the air is evacuated out the cylinder and through the top of the gun. All this happens because of an imbalance of air pressure and flow. When the pressure in the bottom of the body and cylinder is greater it holds the piston driver up. When the trigger is pulled, it creates an imbalance of pressure and flow by exhausting the bottom pressure and diverting the pressure and flow to the top of the cylinder. This drives the piston driver down.

Each different manufacturer build their guns a little different, but the basics of how they work are the same.

In the preventative maintenance section I talked about the things that will wear out a gun. Here, I will explain how to work on your gun. The tools you will need are allen wrenches, screw driver and a pick. A pick is a tool with two different ends. One end is like an old ice pick. The other end is like an ice pick that has the tip bent to a ninety degree angle. This enables you to get the o rings loose from inside of the gun. The moving parts inside the gun have rubber o rings to seal and allow movement. These o rings will wear out and will have to be replaced. If you have to replace one o ring, replace them all. This saves downtime later. They also need lubrication to work properly. Light grease added to the inside of the tool will be made super slick when oil is added on a daily basis.

The gun will be broken down into six areas. They are the cap with poppet, body with cylinder, trigger valve, nose, and magazine. To guide you along you will need the parts schematic. This schematic shows you how the parts are installed and the part numbers for ordering. Look at the schematic carefully. The picture may be upside down to the way you are actually going to install the parts.

Picture #1 below shows the poppet, not numbered, to the right of the piston driver. #2 is the driver and #3 is the piston with the o ring to seal it in the cylinder. You can see the discolored circle around the white poppet, this is where it seals against the cylinder.

Take one area at the time and take it slow. Some manufacturers offer kits to rebuild their guns and some just offer the individual parts.

Picture #1

Some companies offer generic kits on the internet for all brands. Some brands offer trigger cartridges as a replaceable unit and others have to be rebuilt with o rings.

To assess the problem with your gun you will need to listen to the gun when you connect the air hose to it. First would be listening for external leaks. The common areas are the cap, trigger, and nose. You can have a leak at the gasket that seals the cap. Replacing this gasket will fix that problem. If there is a slight warp to the cap or body, it may require two gaskets. This will still allow the gun to function. Any more than that may not allow the poppet to seal at the cylinder. Air leaking from the trigger will require that unit be replaced or rebuilt. The schematic will show which one you have and whether you can rebuild it or replace it as a unit.

Air leaking from the nose will be an o ring around the cylinder that divides the body in half (#2). This particular cylinder skirt has two o rings, inner

Picture #3

and outer. Air leaking out the nose will be air leaking by this partition from the top chamber to the bottom chamber. #1 (pic #3) is the port where the air comes from the handle into the cylinder. #4 is the cylinder that goes from the bottom of the body to the poppet in the cap. #3 is the bottom bumper made of some type of rubber. These will get dry and brittle and need to be replaced at some point.

If air is leaking from the exhaust port in the cap, the poppet o rings are bad. These are on the side of #5 below. The o ring on the top flat surface is a top bumper when the piston driver comes to the top. When you take the gun apart to start checking o rings, check the metal that the o rings touch. This will be #6 in picture #4.

Picture #4

If there was not enough lubrication, or sand
in the gun, there will be a groove in the metal.
If you find a groove in the metal, that part will
have to be replaced. You can try sanding it out
with emory cloth. This may or may not work. If
there is a groove in the body, where the skirt is,
then it will not be cost effective to repair. The
body can be expensive, but you can check the
cost just to make sure. Replacing the o rings
into a worn cap or body will not allow it to seal
even with new o rings. If you have other guns

that you kept for parts, try using one of them. If you are a contractor or carpenter, never throw those worn out guns away. There may be something you can use off of them later.

The internals of the cap has a poppet (#5 in picture #4). Some move more than others. They will be made with plastic, nylon, or metal with rubber o rings. These poppet parts can be broken or cracked. The o rings can be broken or worn. As I stated earlier, if you have to replace one o ring, replace them all. The other parts can be replaced individually if broken. If the cap is worn, they can be replaced at a reasonable cost. This poppet seals the top of the cylinder. Sometimes using emory cloth to smooth out the inside of the metal surfaces will help the o ring seal if there was oxidation or crusting. In picture #4, #3 and #4 are under the poppet. You can see the spring keeps the poppet pushed toward the cylinder for the seal. The white object under the spring is just a cushion and spring holder. #2 in picture #4 is the cap gasket. You can see #1 is a hole that goes from the handle to the top of the poppet. If you put the gasket in wrong and cover this hole the gun will not function. Some gaskets have a hole on each side so you cannot make that mistake, but some don't.

The cylinder should be checked for scratches, sometimes the driver will break and scar the inside of the cylinder. Light scratches can be sanded with emory cloth. If the scratch or dent is too deep, the cylinder and piston won't seal and would need to be replaced. The o rings on the outer parts of the cylinder will also be replaced to seal it in the body. There is an additional o ring or other type of rubber seal to cover some exhaust ports. These ports need to be clean and the seal tight.

The nose leak should be taken care of when the o rings are replaced along with any other broken parts on the cylinder. One problem you will have with any gun is the driver overshooting the nail or staple. In simpler terms, the driver is sliding forward before the nail or staple is fully driven in. You will also see an indention in the wood above the nail or staple. This is caused by a worn nose and or worn driver blade. Replace the driver blade first to see if that corrects the problem because that is more cost effective. Some of these are expensive, but not as expensive as the nose.

The framing nailers are hard to tell when they are worn. They don't seem to wear as bad as the brad and staple guns. The brad nailer, finish nailer, and staplers are easier to see the wear. Open the nose and look for a groove cut in the metal flap that closes against the gun nose. A worn blade and nose has too much movement for the nail or staple to shoot straight. The nail or staple can also get hung up in the nose easier.

When you have an idea of what you will need to replace, make a list of those parts. Then find a place to check the prices and order those parts. When you are doing the work yourself, you will save labor costs and it may be very cost effective to do the repair. If the nose or body of the gun is bad, this may put the repairs cost at almost a new gun's cost. If the gun is old or has been used heavily for some years, it probably is not worth repairing. They do wear out. The magazines are mostly damaged by dropping the gun or getting too dirty. A magazine can cost from thirty to one hundred dollars or more. Sometimes they can be straightened out with a screw driver. There are parts available to replace the ribbon spring that moves the nail pusher. You have to see what is stopping the nail or the pushers movement. If the magazine is broken toward the nose, you may not be able to save it. When you take the magazine off of the nose and handle, watch out for the spring coming loose. You have to make sure that the pusher is moving freely for the nails or staples to advance easily.

Picture #2

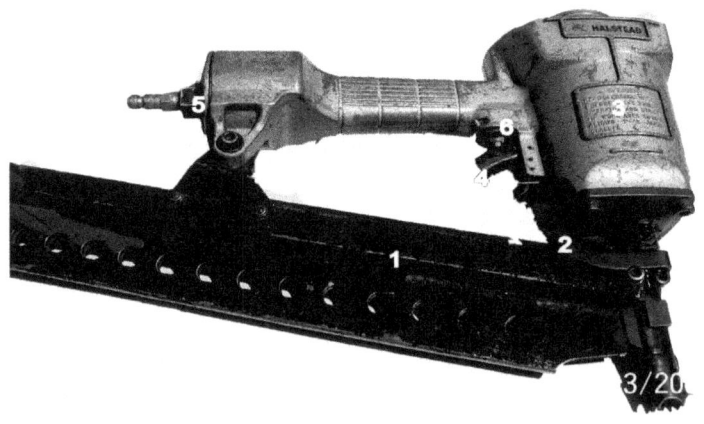

The description of the nail gun body using picture #2 above. #1 is the magazine that holds the nail or staple. #2 is the safety linkage that keeps the nail from shooting unless it is pressed against the wood. This linkage pushing against the trigger valve which activates the valve to shoot the nail. #3 is the body of the gun that holds the cylinder and piston driver. #4 is the trigger that you pull to shoot the nail. #5 is the fitting the air hose connects to to pressure up the gun. #6 is where the trigger valve is housed in the body. It is under the actual trigger lever. These work in combination with the safety linkage. So if you need to remove the trigger valve you will have to remove the trigger and sometimes the safety linkage.

Another problem I have seen is that when a screw is lost that holds the magazine to the nose, it is replaced with one too long. When it is too long it will impede the advance of the nail. If you look down into the nail slot from the bottom of the nose area with a light you can see if the screw is too long or not.

You need to read the owners manual when you purchase an air operated nail gun. Some need to be lubricated and some don't. If certain tools are not to be lubricated, and they are, it will destroy the rubber diaphragm in them. Other brands may not be so drastically affected. Too much oil will not let the gun function, just like too little lubrication will not let it function.

The air pressure is also critical. Too much pressure will blow out gaskets, seals, and cause other internal damage. Too much pressure can even blow the cap off. There is a maximum and a minimum to operate your gun. Some have these limits listed on the gun. If you aren't sure, check your manual or on line for the information. More isn't always better.

Nail gun safety is very important. If you don't think so, do a search on the internet on nail gun accidents. I think that will change your mind. Locking the safeties back is like walking with a loaded shotgun cocked and ready to shoot. They can both kill you or inflict serious damage. Whenever you connect the air hose, always point the nose away from yourself or anyone else. I have had a nail gun cycle just by connecting the air hose. Accidents happen fast, no one wants it to happen. When I am around these tools I always think about what they can do to a hand, arm, face or leg. Some people think they can control the gun under any circumstances, it only takes one time and a split second.

Gas Operated Nail Gun

There are also gas operated nail guns. These guns are great when they work. I do not like working on them. If you have one and it stops shooting, take it to an experienced shop for repair. These are complicated. You have electrical circuit boards, mechanical operation, o rings, gas canisters, and batteries. There are a lot of moving parts in this gun. You will save money and down time by having the work done.

Drills

First thing to do in repairing a drill is to get a schematic and parts list to reference. This will help identify parts as well as put it back together.

There are a lot of different kinds of drills. The motor will be larger or smaller. The chuck end of the drill will be larger or smaller and may have a transmission as well as a gearbox. Then you have AC and DC drills.

AC Drill

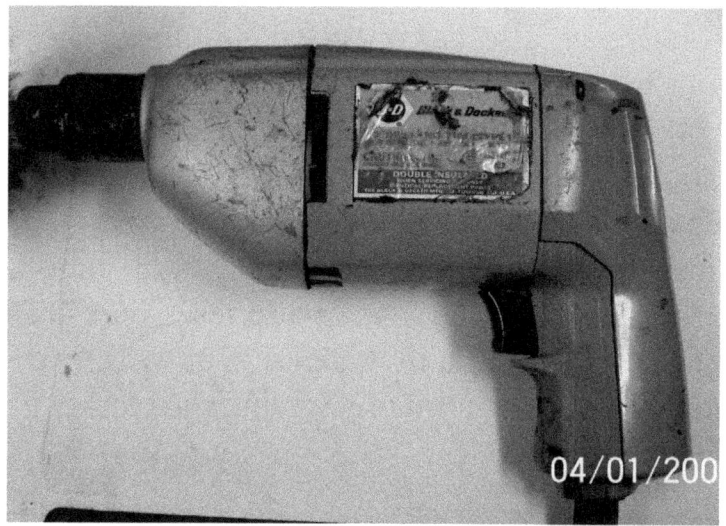

First, the AC drill requires being very careful when working on it. Be careful to unplug it after testing. The first thing to check is the cord for a broken wire or cuts. Use the VOM meter to check the continuity of the wire. If the wire is broken close to the drill handle, you can try cutting it and retesting the wire. If it shows good, you can reuse the cord. If it shows bad, it may be broken toward the plug. In the plug area, if you need a molded water proof plug, you will need a new cord. If the molded plug is not needed, just replace it with any plug that you have. Use the previous chapters on how to use the VOM as a guide in checking your cord.

Picture #1

The cord comes into the drill handle (picture #1) from the right of #1 (pic#1) with #1 being the on/off/speed switch. The white and black wire will go to the switch and the white wire you can see goes to the field in two places. #2 is one of the brushes. The brush makes connection to the commutator on the armature.

The field acts as an electromagnet and the brushes send electric current into the armature. The pulse of electricity in the armature makes it turn. So if you aren't making this circuit the armature won't turn.

When you open the drill to check the cord, check for loose wires. Now use your VOM to check the switch. Most switches have the two cord wires going to one end of the switch and the wires coming out of the switch go to the motor. Place one VOM probe on one end of the switch and the other probe at the opposite end, same side. Then pull the trigger. If you get a reading, that side of the switch is OK. Then check the other side of the switch the same way. If the switch shows good, you will have to move on to checking the field and armature.

Picture #2

Drill parts description picture #2 above. #1 is the gear box that turns the chuck. #2 is the field in which the armature runs. #3 is the fan that moves air through the motor for cooling. #4 is the armature winding. #5 is the commutator where the brushes ride.

There is another method to use in checking a switch. Disconnect the power cord from the switch, then disconnect the wires going to the motor. Make sure your cord is not plugged in, now make a direct connection of the power cord to the wires going to the motor. Make sure the connecting points are taped so they can't touch anything. Now you can plug the cord in. If the motor runs, the switch is bad, if it doesn't, there is another problem.

There are a lot of different kinds of switches in all models and manufacturers. Some have variable speed attachments. Some use the wiring in the switch to reverse the motor. This switch may have quite a number of wires going in and out of it.

To check the field and armature, use the information listed earlier on how to check the field and armature with your VOM.

Picture #3

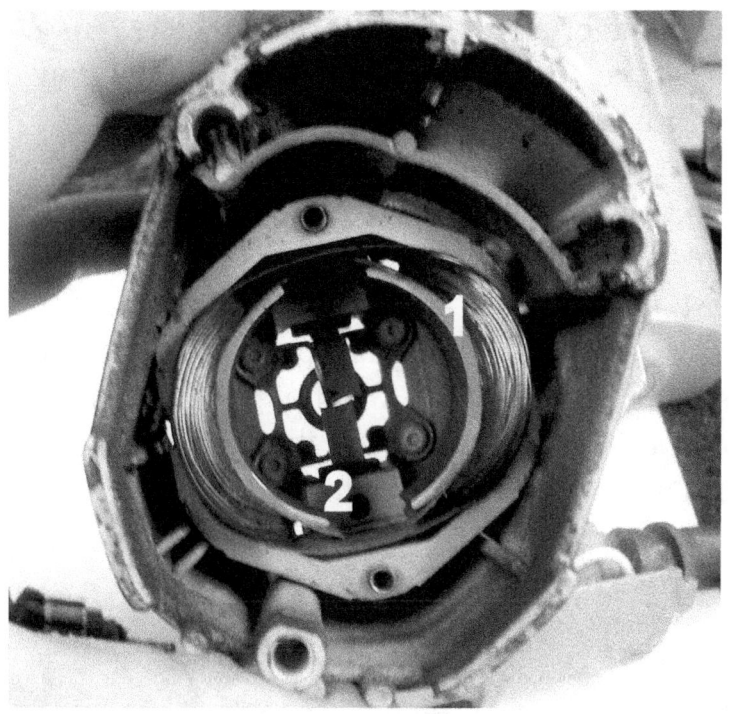

Now you need to check your brushes. Some motors are enclosed and the brushes can't be checked or replaced. Check the area around the brushes for burnt or melted areas. This is from a high temperature and indicates a bad armature or brushes. Loose connections cause excessive heat. Check the length of the brushes. If they are short, replace them. Make sure they move up and down in the brush holder and aren't stuck. If one is stuck, it will not make contact on the commutator of the armature. You can clean the brushes and the brush holders. Also check the springs and the connecting wire on the brush. A broken spring will not push the brush against the commutator enough to make good contact. In the above picture #3 you see what happens if you don't take the brushes out first. #2 in picture #3 is the brush holder. You can see how the spring has pushed both brushes out to the center. You will not get the armature back in unless you remove the brushes so the armature shaft will slide in.

Picture #4

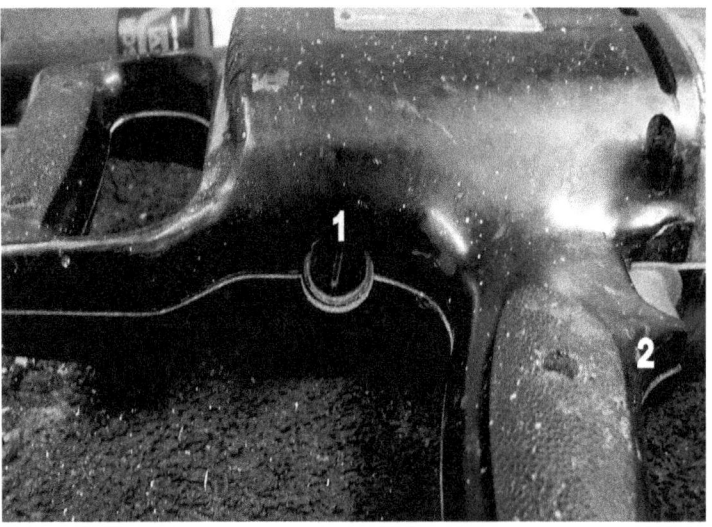

The large drill above (picture #4) is taken apart from the side. #1 is where one of the brushes are removed. The other brush would be on the top of the drill case opposite #1. #2 is the power on/off switch.

Picture #5

In the above picture #5 I will describe the parts you see. #1 is the cap to the brush holder. #2 is motor reverse wiring. #3 is the on off switch body. #4 is the black and white power cord wires going into the on off switch. #5 is the reverse lever.

The below picture #6 is of the drill internals. #1 is the brush holder, #2 is the commutator, #3 is the field, #4 is the fan that moves the air across the motor, and #5 is where the brush touches the commutator.

Picture #6

If the drill is making a lot of noise from the gearbox, the shaft isn't turning, or there is a lot of slack where the chuck moves around, check the gearbox. Some drills have bearings inside on the spindle or shaft. Others may have bushings. If the bearings are bad, you can remove them using the bearing puller and the arbor press. If a gear is torn up, or a shaft is broken, you will have to replace that part. Use your schematic and parts list.

Picture #7

The above picture (#7) is where the gears are housed. Remove three screws (#1) and the back cover comes off to see the gears and lubrication. #2 houses the chuck spindle.

Hammer Drill

The hammer drill is just a larger drill that hammers the bit as it drills. It has a bigger motor and gearbox. This one will have a lot more moving parts. The nose where the bit goes will wear out or get very dirty and not hold the bit. Here again, use your schematic and parts list. Check the lubrication. Is the grease hard in the gearbox? If so, remove it and add a light grease made for tools. A heavy oil is not good to use because it will probably leak out over time.

Battery Operated Drill

Battery operated drills are being used more and more now. The voltages are going higher and the batteries are getting better. Still, people making a living with their drills will wear them out.

Picture #8

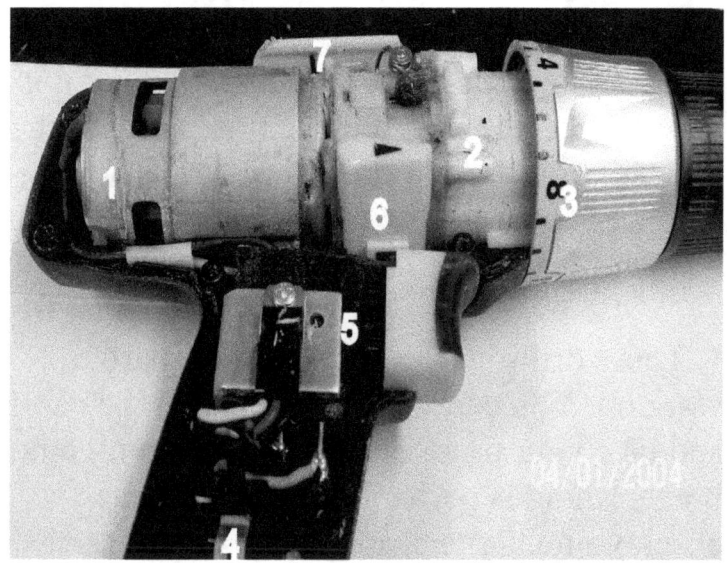

The above battery operated drill (pic #8) is broken down here. #1 is the motor, #2 the gearbox/transmission, #3 the transmission adjustment, #4 the battery contacts, #5 the trigger switch, #6 the forward/reverse lever, and # 7 the Hi and Lo switch for the gearbox.

Most of the problems I see with battery operated drills are using them on the wrong settings. For instance, you have the drill set to drill at high speed and you try screwing in a large lag bolt. Or you have it in third gear when it needs to be in first gear. Think about it like a car transmission, the lower gear has more torque, the higher gear is for speed. Never change gears or settings with the motor is turning. Read the owners manual on how it is to be used. There are a lot of little gears in the transmission and part of it is plastic. They are tough, but you can't try to make it do something it wasn't designed to do.

These drills are complex and are hard to work on. You can work on them, just be careful and take your time. They will only go together if everything is perfectly in place. To start, check your charger with your VOM to see if it is putting out the correct voltage. If it is OK, then check the charge on your battery. Most chargers have a light or multiple lights that give you a signal that the battery is bad or charged.

The batteries are made in different ways. Some have rechargeable C size batteries hooked up in series inside the battery case. If one of those batteries goes bad, you will have a weak battery. If more than one goes bad the battery will not work at all. Some batteries now have electronic boards and chips in them to talk to the battery charger and vice versa. In most cases it is cheaper to purchase a new battery or charger than it is to attempt to repair one.

Once the battery and charger is checked out, now open the gun and look inside. Check the contacts that touch the battery leads and the battery contacts as well. It could be corroded and not making contact. Then look for loose or broken wires.

Now you are ready to check the switch. In the battery operated drill picture the pink and black wire go to the motor from the switch. All the other wires are for the reversing of the motor. You can check it using the VOM as in previous chapters. You can bypass the switch by using two jumper wires. This is a wire with an alligator clip on each end. Attach one to the positive side of the battery and the other end of the wire from the switch to the motor. Then attach the other jumper from the negative side of the battery to the other motor wire. If the motor moves, then the switch is bad. So you have two methods to check the switch.

If the motor still doesn't run, you need to check the armature and field with the VOM as we did in previous chapters. You can see this motor is sealed. There is no way to check anything. If the motor doesn't run when hooked up direct it is bad. The field in some motors will be two solid magnets, not electromagnets. It depends on the motor as to what can be replaced. You need the schematic and parts list to determine this. The more you work with these schematics and compare to your tool, the better you will get at it.

Common Switch Side View

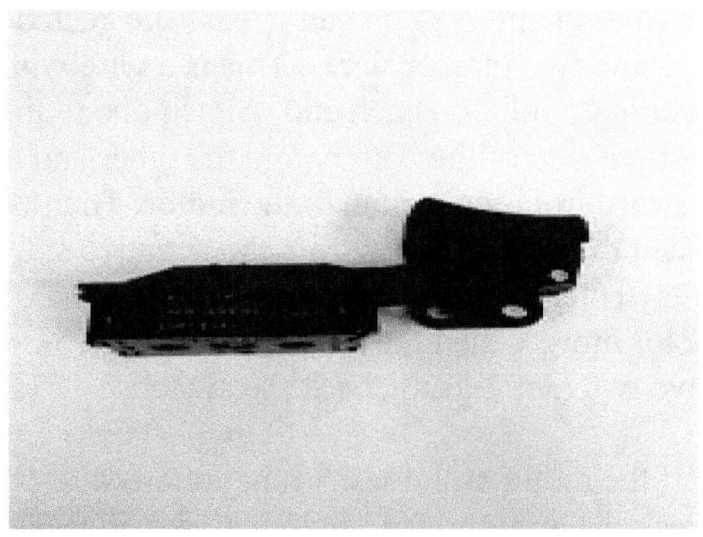

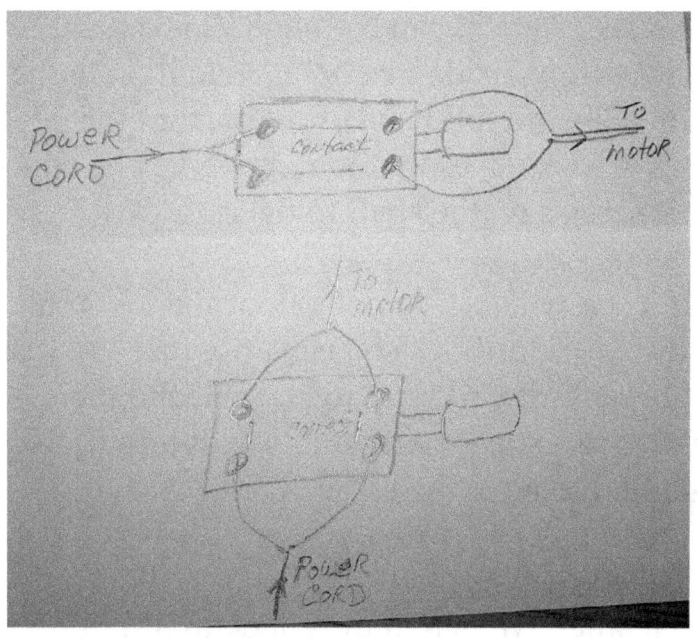

Basic wiring for two different switches used in tools

In the above switch picture I will explain about the wire connections. If you look at the left to right of the switch, the power cord will come in to the left end and from the right end of the switch the wires will go to the motor. Another switch may be connected from side to side of the switch.

Chuck removal can be tough. Some drills have a screw inside that helps hold it on to the shaft. This is a left handed screw, turn the screwdriver to the right to remove it. The chuck itself screws on the shaft to the right.

If a screw doesn't come out or the top of the screw strips, drill it. You can't use just any easy out because it will have to be turned to the right. Use a square easy out, it will turn to the right and remove the screw.

To remove the chuck itself you will have to use an allen wrench in the chuck to turn it off the shaft. Once you have the hex or allen wrench in the chuck, long end out, place the chuck off of the edge of the bench. Use a hammer to hit the allen wrench hard. Keep hitting it until it comes loose. This has not harmed the transmission or gearbox since I have been doing this. If this doesn't work, you will have to cut it off with a cutoff wheel. You have to be careful not to cut into the shaft and mess up the threads. Some people will use a torch to heat the shaft, but there is plastic everywhere so that may cause more harm than good.

I really believe that if you can't get the chuck off easily it is worth the price for a shop to remove it and replace it.

Saws

This can encompass a number of different kinds of saws. These are circular, worm driven, chop, miter, sawzall or reciprocating, and table saws. Then you can add the battery operated.

For these battery operated saws you can use the same methods of checking batteries, chargers, and switches as in the previous chapters. What I will add is that you need to use a sharp blade. The battery operated saws need all the help they can get to do a good job for you.

To work on any of these you need the schematic and parts list to understand how the saw will come apart. If you can't see readily how the tool breaks down, the schematic will show you. Most saws will break down in sections. Be patient.

The AC powered saws will have the electrical end, the gearbox area, blade cover, and the plate the saw rides the board on.

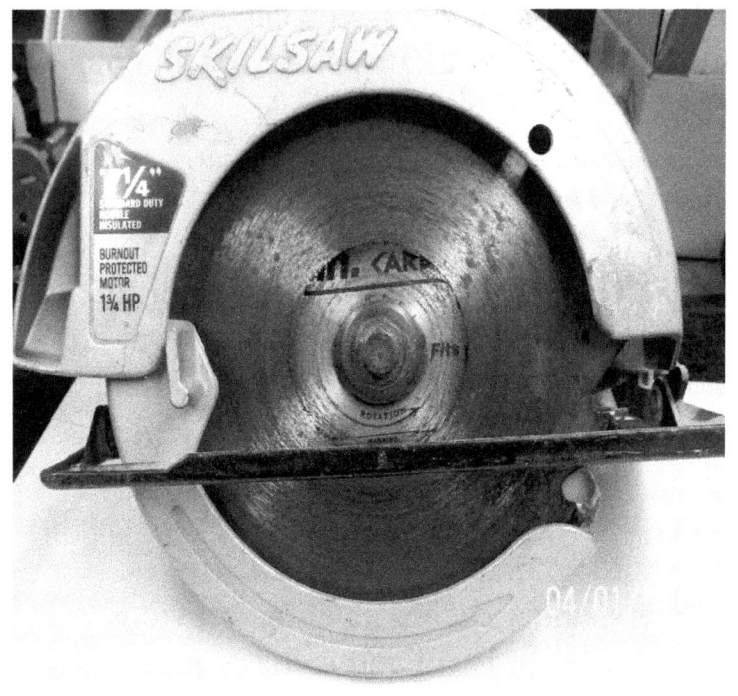

The circular saw gearbox is behind the blade. Remove the blade and the blade guard stop. This allows you to get to all the screws that holds the gearbox entry plate on. Check the grease, add some if you think it needs it. Look at your gears and shaft to see if they are OK. Don't use too much grease. The grease needs to move in the gearbox. Too much grease will cause the motor to be in a strain pulling more amps. This creates more heat, more heat breaks down everything to the point of failure. Use a very light grease that will move around and flow when it gets warm. If the grease doesn't flow it will not lubricate the gears.

If you don't have any light grease, put a small amount of light oil in the gear area. When you get the saw back together, run it for a minute or two. This will hopefully thin the grease so it will move around better.

Before you pull the armature out, remove the brushes. Check the brushes for wear or breakage. You will install the brushes last when putting everything back together.

The electrical end of the saw will be checked as in previous chapters. Check the cord, switch, armature, and field.

The worm driven AC circular saw will have a larger gearbox. If it has run without oil you will need to check the wear to the gears. I have only seen one out of six that didn't need oil in the gearbox. This is a special heavy oil for these saws. Sit the saw level and remove the plug to the gearbox. Add oil until you see it at the bottom of the hole. Install the plug. Any bearing you can access should be checked; do they spin freely and are they tight? Again, use your schematic to guide you.

In the above picture is what I use when the saw has no blade lock. Put a screw driver in the teeth of the blade to hold the blade and use the wrench to remove the bolt that holds the blade on. The bolt will loosen to the left.

To remove the plate to access the gears. Remove #2, the blade guard stop, and then the small screws behind #3. Then you can see the gears and lubrication.

Sawzall or Reciprocating Saw

The sawzall or reciprocating saw electrical is checked the same as in previous chapters. This gearbox will be different, but will need to have the moving parts checked. Check the lubrication and add if necessary. Gears and bearings are important to check as well.

Sometimes it is obvious what is wrong and sometimes you have to really look closely. These can be very tricky to put back together.

The blade holders on a reciprocating saw can be from very complex to very simple. Sometimes all they need is lubricating and cleaning. If something is broken, you will need your schematic. Some blade holders push and twist. These may lock open and will not release until the blade is inserted. Then it will close.

The chop and miter saws are similar. The electrical checks on these are the same as in previous chapters. They are basically circular saws on a base with a pivot. Again, get your schematic out to see how it comes apart to check the position of the moving parts. There will be bearings on the armature shaft. These go out quite frequently. I believe it is for a couple of reasons. The bearings themselves may not be a good quality and the belt from the motor shaft to the blade shaft is too tight. Use the bearing puller and arbor press to change these bearings as explained in a previous chapter.

The table saw will be made a little different by different manufacturers. Some may have the blade on the motor shaft. Others will have a spindle with the blade attached turned by a belt and motor. The schematic will show you how it comes apart and how it goes back together. The electrical will be checked as in previous chapters. Check the bearings on the shafts and the armature. Most of these saws that make noise will have bad bearings or overheat causing damage to the armature and field. Where there is evidence of high heat you will need to replace the armature, field, and brushes. If you don't, it will go down again.

Be sure to blow out all motors before reassembling. This will help air flow to remove more heat. You can also blow the motor out if it is not taken apart to remove the sawdust and increase air flow. This will be the primary cause of any motor overheating. Loose connections will also cause heat issues.

There are switches that are used with tools that have breaks. Replacing the switch with the part called for on the parts list will ensure that you get the proper switch. Generic switches are sold and can be wrongly sold as a replacement.

If the break doesn't enter into the picture, then it will not matter. When you go to a tool store for parts, you assume they know the proper switch to sell you. If you are informed about your tool and have the correct information and model number, you will get the correct parts. You can also bring the old part with you to compare.

If you replace a switch and the wires are connected end to end and the tool still doesn't work, try a side to side connection. The connections on some switches are obvious and some aren't.

Sanders

There are a lot of different sanders, but all will be checked basically the same. There are palm and orbital sanders of different sizes and horse power. Then you have the portable and stationary belt sanders. When looking at the electrical part of a sander, we will check it as we have all the rest of the AC tools. The switches on these will be smaller. Some will be a trigger switch and others just an on off toggle.

Some of these sanders will have bearings and some will have bushings. The schematic will show you what to look at. Most of the problems in sanders are bearing and motors burning up. This tool runs a lot longer than most tools and have the fine dust to suck up in the motor. Brushes wear out quicker also. Blowing out the motor regularly will help extend its life.

The portable and bench top belt sanders are a little different. These will have belts to drive the sanding belt. Some of the rollers may have bearings or bushings. The schematic will identify this. Noise when sanding will be the key here. The electrical should be easy for you to check now.

Some of the lubrication used in sanders will be a light machine oil. The more you use it, the more you need to blow it off and lubricate it. Most bearings used are sealed and will not accept any lubrication. If it has bushings, these can be lubricated.

Some sanders will have a belt or something that looks like a rubber band. If the motor has brushes you need to remove them first to take the rest of it apart. The movement of the brushes is important as well as the length of the brushes.

If you look at the schematic you can see if there are gears, belts, bearings or bushings. You will have an idea of what needs to be lubricated and parts that can break to stop it from working.

Hydraulic/Air Rivet Gun

This is a tool not everyone will have in their shop. You may be that person that uses one, so here is the info. For some reason this tool uses both air and oil to operate it. I guess you could say it is hydraulic.

When you take the tool apart, the top portion will unscrew from the bottom cylinder. Oil will spill out. The cylinder may be full or close to it. There are o rings in some of these moving parts that need to be checked. Most of the problem with these tools will be the oil level in the cylinder.

When the air enters the the bottom of the main cylinder, as the piston moves up, it will push the oil in the cylinder up. The oil will enter the head that will push against a spring, making the fingers holding the rivet retract. When the air is shut off, the spring pushes the oil back into the cylinder and the fingers are ready to retract the rivet again.

When the gun stops working you will have to check the oil level. The shaft should be pushed down so the piston is bottomed out. On the opposite side of the air chamber, is an oil level. With the piston retracted to the bottom of the cylinder, it is ready to add the oil. Tool oil can be used to fill the cylinder. Take the top of the rivet gun and screw it back on the base. Only screw it down a little and use the lock ring to lock the two sections to keep them from moving. What this does is allow you to adjust the lower part of the gun up to compress the oil and keep the gun working. You will make this adjustment when the rivet gun gets sluggish or is not working. Once the adjustment reaches the end of the threaded area, more fluid will need to be added. If you see noticeable leakage, this will be the problem and should be repaired. There may be a slight leak internally causing the fluid loss into the air chamber.

There are three grippers in the nose that will grip the shaft of the rivet. When you pull the trigger, the gripper grabs the shaft and pulls the rivet snug in the hole. You have to pull the trigger again for the gripper to pull the head through the rivet and cut the shaft off.

If you can find a schematic on your rivet gun, this will help you check the moving parts to see what they should look like. If the oil level doesn't fix the problem I would change out all the seals inside.

Air Impact and Ratchet

Most of what I have seen with these tools are not enough lubrication or water in the gun. They will have dirt and rust in them. Most of the time a good cleaning and lubrication will fix them. There are a lot of different models and configurations. The basic operation for all are the same. Air going in turns a shaft with vanes and then exits.

Find a schematic for your tool so when you break it down you can put it back together correctly. If you can't find one, take it slow.

You will need a rag, emory cloth, and penetrating oil. Take it slow.

Some of the guns have metal veins that slide around inside a cylinder. Be careful because these veins will fall out when you open up the gun.

The best way to work on a tool is to have it on a surface that is soft, like a rag. Get all your tools together, emory cloth and lubricant. Try to do it all the same day so you won't forget how it went together. Most of these tools are simple. The cheaper models will be hard to find parts for. The name brands and better models will be easier to locate parts. The cheap models are just that, cheap for a reason. They want you to throw them away. They will not offer parts for some models, or only have parts available for a couple of years.

Drain your air compressor tanks regularly. If the gun is still working, run oil through it until it is running better. This may get it working without having to take it apart. I would still use tool oil so no o rings will be affected by using the wrong oil. If you do use a penetrating oil, follow it up with tool oil.

Electric Impact Gun

The electrical end of this tool will be checked as all the rest of the electric tools in previous chapters whether AC or DC. Check your cord, switch, field, battery, charger, field, and armature. You need to blow this motor out too. The hotter the motor gets the more it breaks down. Check and clean the brushes and brush holder as talked about previously.

You will need your schematic to break open the gearbox. When inside, check for worn gears or broken shafts. Check for good lubrication also. Do not overfill the gearbox with grease, it will force itself out somewhere. There needs to be room for it to move around. If there are bearings, check them also.

If the anvil is broken, it will probably not be worth fixing. The anvil is the nose of the tool that the socket attaches to. It is the most expensive part of the tool.

Chapter 4
Tools in General

Cheap tools are just that. Most of them are not made to be rebuilt. You may have a name brand, but does that model exist on parts sites? The big box stores, as they are called, will have models made just for them by a name brand tool manufacturer. They will be made in such a way as to be able to sell the tool cheap. You may be able to buy some parts or none at all.

Some tool manufacturers will not sell parts to an individual or a repair shop. You have to send the tool in to their service centers. Ask about tool repair when you are looking at a tool. Do some research.

Some tool parts, depending on the manufacturers, will be expensive. Other tool companies will have better pricing on their parts. Some model tools will have parts that wear out over time and are so expensive to replace that it would not be cost effective to fix it. Some different manufacturers parts are interchangeable. When you look at a nail gun and it kinda looks like another brand, chances are some of those parts will be interchangeable.

There are some generic after market companies selling parts for name brand tools. You have to surf the internet to find these companies.

Ask questions in a tool store that knows something about tools. The big warehouse stores can't answer a question except maybe what the price is. Your independent tool dealers will answer questions, repair, sell you parts, and give you information. They will have accessories you need and could be a dealer that handles warranty work. I believe the independent tool dealer, with what they can give you, is a good value. They can recommend the different nails your gun can use by name or generic brand to save money.

I believe in attempting to repair things, but if I can't, then I will pay for someone to fix it. So, worse case, put it in a box and take it to the shop.

There is a lot of information on the internet. You can find electrical diagrams, tool schematics, parts stores, forums on how things work, and much more. Be informed on your tool or research it before you go buy it. You may look at the schematic and see the motor doesn't have brushes and has an enclosed motor. This motor may cost double what a motor with brushes costs. You may find out that you can't even find that part number listed by any parts suppliers. That would be costly. Research more than just the price of the tool.

Chapter 5
Safety

I have mentioned about some safety factors in the previous chapters. Even though I am talking about safety last doesn't mean it is not important. Anytime you can keep all your fingers or keep from getting shocked, it is important.

The most obvious safety subject may be electricity. There are tools out there that use 110v and 220v AC. Some of these tools have breakers, but that is to save the tool, and may not save you. You can buy an extension cord that has a ground fault breaker or an outlet that is a ground fault itself. The ground fault is suppose to break the current flow before it can shock you. The one problem with ground faults are that they are a weak point and have to be replaced more frequently. This is better than getting shocked.

When running extension cords across a large area, you have to be mindful of water puddles and the condition of your cord. Most cords have molded plugs that are water tight. Most tools now have a lot of plastic in them or have plastic handles. If you are holding the handle and a wire inside comes loose and touches the handle, you are OK. If the handle is metal and no ground fault, you are depending on the 110v electrical breaker to keep you alive.

I see cords cut into by circular saws all the time. This is where the plastic handle and ground fault pay for themselves. You should always use wire nuts or tape when tying wires together inside the tool. This helps hold the wires together, but will also protect you from touching a live wire. This may be important if you forget to unplug the tool after testing and open the tool up to work on it.

Blade guards are very important and still people will wedge them open with a piece of wood. This will not allow the guard to close. This can cause a number of problems. If you set it down, it will run off and ruin the blade or any number of other things. Keep your guards on all your tools in working order. They are your first line of defense with a hand held saw.

Most accidents will happen when you are doing very repetitive work or you get too comfortable in what you are doing. Whenever you start a project or start your day on the job, think about safety. It is such a little thing to do to get such a big payoff. Going home that day with all your fingers and your eye site should be high on your to do list for the day.

Some people wear glasses all the time and that will help you. If you don't, get some safety glasses, goggles, or face shield. Things happen so fast that you can't close your eyes, so be prepared.

There is another practice that could turn deadly that a lot of people probably don't think about. Compressed air. If you direct a high pressure air stream onto your skin and it penetrates, you could have an air bubble moving in your blood stream. If the bubble goes to your heart it may cause a serious problem or even death. This is not to mention that air at a high pressure may tear or cut your skin badly.

One item that has not been discussed in this repair guide because it really doesn't apply, is a ladder. The ladder certainly is an item that would be used with these tools. I think most of the time when we are using a tool to do something we may have a thought that says, " This isn't safe." It probably isn't, why take the chance? You don't have to fall far to fall hard. Don't go too high or lean too far side to side. Make sure you are on stable ground. If you have to go high, it would be good to have someone help to hold the ladder or hand you things. Think through what you are doing, no matter what it is, that will help.

When we look at air compressors and generators there are a number of things to look for. Belt guards are important for protection. Some mufflers have guards, but are still hot. Exhausts can cause sparks if they don't have an arrestor on the muffler. Don't run a motor operated compressor or generator inside a closed area with no ventilation. The carbon monoxide will make you sick or kill you. Be very careful when refilling the gas tank that there is nothing hot enough around to ignite the vapors. No Smoking around a tool's gas tank or gasoline can.

Some people like to disable safeties on whatever tool they are using. This can be deadly for you or someone close to you. I equate a nail gun as a 44 magnum pistol. I always point it away from myself or anyone when plugging the air hose up to it. I have had them fire when plugging the air hose to the gun. You can inadvertently touch the trigger and the gun fires. Will it fire a nail at you or someone else?

Watch where you put your hand or fingers when cutting a board or shooting a nail or brad. If you are shooting a nail at an angle it could glance off the board. Position yourself for this possibility. Safety glasses would also help.

There are a lot of different tools used by homeowners and construction personnel. We will never be able to go over every tool and what to watch out for. You need to look at the tool, what it does, and make yourself aware of the possibilities when using it. Sometimes we may get a second chance at something, but sometimes we won't. The more we gamble, the more likely we are to lose. Be Careful!

Epilog

I hope I have given you enough information to be able to make some tool repairs and save some money. If you save one tool, make one repair or help someone else make a repair, you have more than paid for this book. The more you work at tool repair the better you will get at it.

If you have a tool that is not worth fixing, tear it down and use it to learn. Friends may have old tools that are no longer of any use, take them and practice.

I would like to touch on safety again as I bring the book to a close. I try to always think about where my hands are when using a tool. Respect the tool and what it can do every time you pick one up. If safety is on your mind, you will probably be OK. There is a time and place for all safety gear. It is a hassle, but can you close your eye in a split second? Help others think about safety as well. It could be their nail gun that has the safety locked back and shoots a nail at your head.

Last but not least is to thank my wife Barbara and my daughter Kellie for their help with proof reading and solving whatever problems I had. It really helps to have other eyes looking to see if it makes sense. It is also nice to have support in whatever you are attempting to do.

I hope you are successful in all your tool repair attempts. Remember how important preventive maintenance can be to your tools.

The End

www.ingramcontent.com/pod-product-compliance
Lightning Source LLC
Chambersburg PA
CBHW051709170526
45167CB00002B/600